Regression for Economics

Regression for Economics

Shahdad Naghshpour

businessexpert
Press

Regression for Economics

First published in 2012 by
Business Expert Press, LLC
222 East 46th Street, New York, NY 10017
www.businessexpertpress.com

ISBN-13: 978-1-60649-405-9 (paperback)

ISBN-13: 978-1-60649-406-6 (e-book)

DOI 10.4128/9781606494066

Business Expert Press Economics and Finance collection

Collection ISSN: 2163-761X (print)
Collection ISSN: 2163-7628 (electronic)

Cover design by Jonathan Pennell
Interior design by Exeter Premedia Services Private Ltd., Chennai, India

First edition: 2012

10 9 8 7 6 5 4 3 2 1

Printed in the United States of America.

To Parisa
SN

Abstract

The concept of regression was introduced by Sir Francis Galton, but R.A. Fisher provided the statistical theory and application for it for the first time. The 20th century witnessed the spread of regression analysis into every scientific branch. Regression analysis is the most commonly used statistical method in the world. It is used in economics and many other fields. Although few would characterize this technique as simple, regression is in fact both simple and elegant. The complexity that many attribute to regression analysis is often a reflection of their lack of familiarity with the language of mathematics. But regression analysis can be understood even without a mastery of sophisticated mathematical concepts. This book provides the foundation of the regression analysis. All the examples are from economics, and in almost all the examples the real data is used to show the applications of the method.

This book seeks to demystify regression analysis. The concepts related to regression analysis are explained in a way that is comprehensible to those whose mathematical skills are not expert. There is logic to regression analysis that resembles the intrinsic logic that we apply in comprehending the various events that fill our lives, which are probabilistic rather than deterministic in nature. What hinders peoples' comprehension of regression analysis is the difficulty many have in understanding mathematical symbols and derivations. By removing this obstacle, this book enables the logical reader to learn regression without possessing superior mathematical skills. Although this proposed book will be largely nonmathematical in its approach, it will not in any way give short shrift to the subject of regression. This book is targeted to all business students and executives who need to understand the concept of regression for practical and professional purposes.

The regression analysis can be used to establish causal relationship between factors and the response variable. However, in order to be able to do it, the economic theory must be used to provide causal relationship and apply the regression analysis to verify the validity of the theory.

This book utilizes Microsoft Excel to obtain regression results. Although spreadsheet software is not the software of choice for performing sophisticated regression analysis, it is widely available. Moreover, the use of Excel will preempt the need to buy and learn new software; in itself another impediment to learning and using regression analysis.

Keywords

regression, analysis, causality, inference

Contents

Foreword

Statistics Is the Science of Finding Order in Chaos

Regression analysis is by far the most commonly used statistical analysis tool in many areas of science, including Economics. After you finish the book, I hope you will agree with me that if there was one tool tailor-made for economics, it must be regression analysis. They are many aspects of regression that perfectly match the needs of an economist.

Often students of introductory statistics are overwhelmed because of the diversity of the material. There are too many new concepts and too many different topics, which may not seem related in any sensible way. In regression analysis, the focus is on one and only one topic, regression analysis. This narrow focus is due to several reasons. Reason one is that after having been exposed to introductory statistics, you are now ready to focus on a special topic. Reason two is that the topic is so vast that even dedicated books are sufficient to cover all aspects of the topic. The present manuscript does not even scratch the surface of the vast topic of regression analysis. My hope is that you learn to see economics from an applied angle and manage to focus on specific outcomes and their magnitude. I want you to know that every claim in economics is a testable hypothesis, and every theorem in economics can be written as a regression model and thus tested for the magnitude of the expected outcome. Regression analysis or its broader subject area, statistics, is not a substitute for economic theory. Instead, it is a complementary tool that allows us to estimate the magnitude of the theoretically predicted outcome and to test the results against the claims of policy makers and planners.

Acknowledgments

I am indebted to my wife Donna who has helped me in more ways than imaginable. I do not think I can thank her enough. I would like to thank Michael Webb for his relentless assistance in all aspects of the book. He has been my most reliable source and I could always count on him. I also want to thank my graduate assistants Issam Abu-Ghallous and Brian Carriere. They have provided many hours of help with all aspects of the process. Without the help of Mike, Issam, and Brian, the book would not have been completed. I also would like to thank Madeline Gillette, Anthony Calandrillo, and Matt Orzechowski who read parts of the manuscript.

Introduction

Economics is a very interesting subject. The scope of economic domain is vast. Economics deals with market structure, consumer behavior, investment, growth, fiscal policy, monetary policy, the roles of the bank, etc. The list can go on for quite some time. It also predicts how economic agents behave in response to changes in economic and noneconomic factors such as price, income, political party, stability, and so on. The economic theory, however, is not specific. For example, the theory proves that when the price of a good increases the quantity supplied increases, provided all the other pertinent factors remain constant, which is also known as *ceteris paribus*. What the theory does not and cannot state is how much the quantity increases for a given increase in price. The answer to this question seems to be more interesting to most people than the fact that the quantity will increase as a result of an increase in price. The truth is that the theory that explains the above relationship is important for economists. For the rest of the population, the knowledge of that relationship is worthless if the magnitude is unknown. Assume for 10% increase in price the quantity increases by 1%. This has many different consequences than if the quantity increases by 10%, and totally different consequences if the quantity increases by 20%. The knowledge of the magnitude of change is as important, if not more important, than the knowledge of the direction of change. In other words, predictions are valuable when they are specific.

Statistics is the science that can answer specific issues raised above. The science of statistics provides the necessary theories that can provide the foundation for answering such specific questions. Statistics theory indicates the necessary conditions to set up the study and collect data. It provides the means to analyze and clarify the meaning of the findings. It also provides the foundation to explain the meaning of the finding using statistical inference.

In order to be able to make an economic decision, it is necessary to know the economic conditions. This is true for all economic agents, from the smallest to the largest. The smallest economic agent might be an individual with little earning and disposable income, while the largest can be a multinational corporation with thousands of employees, not to mention governments. Briefly, we will discuss some of the main needs and uses of statistics in economics and then present some uses of regression analysis in economics as well.

The first step in making any economic decision is to gain knowledge of the state of economy. Economic condition is always in a state of flux. Sometimes it seems that we are not very concerned with mundane economic basics. For example, we may not try to forecast what the price of a loaf of bread is or a pound of meat. We know the average prices for these items; we consume them on a regular basis and will continue doing so as long as nothing drastic happens. However, if you were to buy a new car you would most likely call around and check some showrooms to learn about available features and prices because we tend not to have up-to-date information on big-ticket items or goods and services that we do not purchase regularly. The process described above is a kind of sampling, and the information that you obtain is called **sample statistics**, which you use to make an informed decision about the average price of an automobile. When the process is performed according to restrict and formal statistical methods, it is called **statistical inference**. The specific sample statistics is called sample mean. **Mean** is one of numerous statistical measures at the disposal of modern economists. Another useful measure is the median. The **median** is a value that divides observations into two equal halves, one with values less than the median and the other with values more than median. Statistics explains when each measure should be used and what determines which one is the appropriate measure. Median is the appropriate measure when dealing with home prices or income. Applications of statistical analysis in economics are vast, and sometimes they reach to other disciplines that need economics for assistance. For example, when we need to build a bridge to meet economic, social, and even cultural needs of a community, it is important to find a reliable estimate of the necessary capacity of the bridge. Statistics indicates the appropriate measure to be used by teaching us whether we should use the median or the mode. It also provides insight on the role that variance plays in this problem. In addition to identifying the appropriate tools for the task on hand, statistics also provides the methods of obtaining suitable data and procedure for performing analysis to deliver the necessary inference.

One cannot imagine an economic problem that does not depend on statistical analysis. Every year, the Government Printing Office compiles the Economic Report of the President. Although the majority of the statistics in the report are fact-based information about different aspects of economics, many of the statistics are based on some statistical analysis, albeit descriptive statistics. **Descriptive statistics** provides simple yet powerful insight to economic agents and enable them to make more informed decisions.

Another component of statistical analysis is inferential statistics. **Inferential statistics** allows the economist and political leaders to test hypotheses about economic condition. For example, in the presence of inflation, the Federal Reserve Board of Governors may choose to reduce money supply to cool down the economy and slow down the pace of inflation. The knowledge of how much to reduce the supply of money is not only based on economic theory, but also depends on proper estimation of the final outcome.

Another widely used application of statistical analysis is in policy decision. We hear a lot about the erosion of the middle class or that the middle class pays a larger percentage of its income in taxes than the lower and upper classes. However, how do we know who is the middle class. A set dollar amount of income would be inadequate because of inflation, although, we must admit even a single dollar amount must also be obtained using statistics. However, statistical analysis has a much more meaningful and more elegant solution. The concept of interquartile range identifies the middle 50% of the population or income. Although interquartile range was not designed to identify the middle 50% and is not explained in these terms, the combination of economics and statistics is used to identify the middle 50% for economics and policy decision purposes.

The knowledge of statistics can also help to identify and comprehend daily news and events. Recently, a report indicated that the chance of accident for teenage drivers increases by 40% when there are passengers in the car that are under 21 years of age. This is a meaningless report. Few teenagers drive alone or have passengers over 21 years of age. Total miles driven by teenagers when there passengers under 21 years of age far exceeds any other types of teenage driving. Other things equal, the more you drive, the higher the probability of an accident. This example indicates that the knowledge of statistics is helpful in understanding everyday events and in making sound analysis.

When an economic phenomenon is changed to produce a desirable income, we need more powerful tools than simple statistics. **Regression analysis** is one of the most widely used statistical tools at the disposal of economists.

In regression analysis, the effect of one or more factor is measured to determine another factor. The first group is also known as explanatory variables, while the latter is known as endogenous variables. In economics it makes sense to refer to explanatory variables as policy instruments. Policy instruments are variables that economists and policy makers can change or control. The supply of money is a policy instrument controlled by the Federal Reserve. The Fed has to collect data first, which is done on a periodic basis. These statistics inform the Fed that there is a problem in the economy, such as inflation. The Fed decides to reduce the supply of money. It will wait for the economy to respond to the change in supply of money. Then economic indicators are measured again and tested against the target set by the policy. If the policy objectives are not met, the action is repeated until the desirable outcome is obtained.

When working with a regression model, one might wonder if it was designed to serve economists. Even some of the commonly used terminologies are the same in both fields. For example, both subjects use explanatory variables to measure the response variable. Typical regression models do not consist of one explanatory variable and one response variable. Instead, in addition to explanatory variables, the model has additional variables known as control variables. Control variables are actually the same thing as economics shifters. **Shifters** in economics refer to variables that are assumed to remain constant for the sake of identifying the impact of the explanatory variables on the response variables. In fact, every economic theory seems to have the famous *ceteris paribus*, which means other things being equal. When other things are not equal and change, they do not distort the relationship between explanatory and response variables. They simply shift the magnitude up or down, depending on the direction of the impact. Estimation of demand provides a good example. Economic theory states that an increase in price reduces the quantity demanded, *ceteris paribus*. The regression model for this economic theory can be written as

$$Q_d = \beta_0 + \beta_1 P + \varepsilon \tag{I.1}$$

where ε is the error term, which will be explained later. To complete the process, we need to test the hypothesis that the coefficient of price, which is also the slope of the demand curve, is negative. So we use statistics to test the following hypothesis:

$$H_0: \beta_1 = 0 \qquad H_1: \beta_1 < 0$$

The model, however, is not complete, because it is not subject to *ceteris paribus* as it does not control anything. Simple control variables consist of price of a complementary good, a substitute good, and income, to name just a few important ones. The theory predicts that the effect of a change in the price of a complementary good is inverse, the effect of a change in the price of a substitute good is direct, and the effect of change in income is direct. Thus, model (I.1) should be modified as below.

$$Q_d = \beta_0 + \beta_1 P + \beta_2 P_c + \beta_3 P_s + \beta_4 Y + \quad + \varepsilon, \tag{I.2}$$

The theoretical claims are written as

$$
\begin{aligned}
H_0: \beta_1 = 0 \qquad & H_1: \beta_1 < 0 \\
H_0: \beta_2 = 0 \qquad & H_1: \beta_2 > 0 \\
H_0: \beta_3 = 0 \qquad & H_1: \beta_3 < 0,
\end{aligned}
$$

where the subscripts use the first letters of complementary and substitute, and Y represents income. The regression model clearly and perfectly matches the economic theory from expected effects of each variable to the concept of *ceteris paribus*.

CHAPTER 1

The Concept of Regression

Relationship Between Variables

Often we are interested in explaining a phenomenon using other factors. There are numerous methods for accomplishing this objective. When the phenomenon is quantitatively measurable, the solution is much easier and the methods are well established. One such method is **regression**.

In regression analysis, one variable (**dependent variable**) is explained by one or more variables (**independent variables**). Before explaining a regression model, presenting an example of a simple model for explaining consumption using income is beneficial. But we first need to define the economic concept **marginal propensity to consume (MPC)**.

Definition 1.1

The **marginal propensity to consume or MPC** represents the amount one would consume if one is given an extra dollar.

$$\text{Consumption} = \text{subsistence consumption} + \\ (\text{marginal propensity to consume}) \times (\text{income}). \tag{1.1}$$

Conceptually, MPC is the same as the slope of regression line when there is only one independent variable. In equation (1.1), **consumption** is the dependent variable and **income** is the independent variable. Although the term **dependent variable** is commonly used in economics literature, other names such as **endogenous variable**, *Y* **variable**, **response variable**, or even **output** are often used as well. Similarly, the term **independent variable** might be replaced by **exogenous variable**, *X* **variable**, **regressor**, **input**, **factor**, or **predictor variable**.

Equation (1.1) is a good example of the concept of regression, but it is not a regression model. The format for a regression model will be discussed shortly. You are more likely to be familiar with a mathematical function than a statistical function such as regression. A mathematical function represents a nonprobabilistic association between a dependent variable and one or more independent variables; the association is exact and fixed (Figure 1.1a). A regression model is a simplification of reality. It is actually a **claim** of a relationship and thus, a **testable hypothesis**. The association between the dependent variable and the independent variable(s) is **probabilistic** and not deterministic. It is true on the average only. Figure 1.1b depicts pairs of (X, Y) observations relating dependent variable (Y) to the independent variable (X). Many factors affect the actual value of Y and cause the observation to deviate from the expected values. A regression model represents the expected value.

Equation (1.1) is the equation of a line except that it is not written in the customary form (used in geometry). It is also a "function" because it provides a specific outcome based on a linear rule, that is, as income changes, consumption changes by the magnitude of the *MPC*. If income becomes zero, consumption drops to the level of subsistence consumption, which is the level of consumption necessary to survive even if one does not have any income. Note that here we are not interested in answering how one manages to pay for subsistence consumption, which could be from savings, selling household furniture, or something else. That is

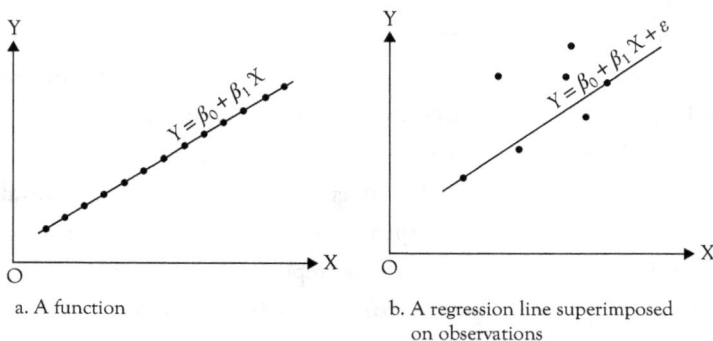

a. A function

b. A regression line superimposed on observations

Figure 1.1. Comparison of (a) a function with (b) a regression model.

not the purpose of this model. The purpose is to explain the level of consumption in response to changes in income. This model is a simplification of reality. For example, it does not take into account the role that wealth might play in explaining consumption. In a more elaborate model, additional independent variables could be included that might improve the model's ability to estimate the dependent variable more accurately and to more closely approximate the reality.

Although this model is a good starting point, it is not a precise replication of reality. Nevertheless, it is the same as a simple consumption function explained in many introductory macroeconomics textbooks. As such, it serves a similar purpose: introduces the concept, clarifies application of the concept, and prepares for a more appropriate model.

Definition 1.2

A *model* is a simple representation of something real in life.

The level of representativeness is determined by the purpose of the model and does not necessarily make a model more desirable, in part because the purposes of a study affect the desirability of the level of sophistication of the model.

Models need restrictions on their parameters to make sense. For example, the *MPC* has to be positive and less than one. A negative *MPC* means that as income increases, consumption decreases and eventually drops below subsistence level, while an *MPC* greater than one means that consumption at some point becomes larger than income. *MPC* values below zero or above one contradict reality and defy common sense. Therefore, we restrict *MPC* to be between 0 and 1. In addition, negative values for the independent variable of income and the dependent variable of consumption are meaningless. Similarly, a negative subsistence level would be impossible. However, there are situations where the estimate for the subsistence level might turn out to be negative, but for the purpose of this example they can be ignored.

The four values of income, consumption, the *MPC*, and the subsistence level are very different from each other. Consumption and income, the dependent and independent variables, are observable data. This means we can gather data on actual income and consumption

levels of a sample of people. The data are typically published and customarily represented in a column format. Subsistence consumption
and *MPC*, however, are known as **parameters**. Parameters are almost
always unknown and have to be estimated. Although every nation has
an *MPC* at any given point in time, the actual value is unknown, as
is the case with the subsistence level of consumption. The parameters
are estimated by the model using regression analysis. In the jargons of
regression, parameters are sometimes called **coefficients** or **slopes**. The
interpretation of coefficients and their appropriate analyses are covered
in Chapter 6.

Definition 1.3

A *parameter* is a characteristic of a population that is of interest.
Parameters are constant and usually unknown.

Examples of parameters include population mean, population variance, and regression coefficients. One of the main purposes of statistics
is to obtain information from a sample that can be used to make inferences about population parameters. The estimated value obtained from a
sample is called a **statistic**.

Definition 1.4

A *statistic* is a numerical value calculated from a sample that is variable
and known.

The word **statistic** has several meanings depending on the context:
two of its meanings are presented in the previous paragraph. The first use
of the word refers to the science and the discipline of statistics. The second
use is more specific and is based on the above definition. In the science of
statistics, we use **statistics** to make inference about **parameters**.

The slope and intercept terminologies used in geometry are also
commonly used to refer to coefficients in regression analysis. In the
consumption model, the corresponding analogy to geometry is that
MPC is the **slope** and subsistence level is the intercept of the consumption line. According to this model, a dollar increase in income increases
consumption by the magnitude of *MPC*, which by definition is the slope

of regression line. When income is zero, the amount of consumption is equal to subsistence level and therefore, indicates the intercept.

The representative terms consumption and income used in equation (1.1) only apply to this particular problem, which renders them inapplicable when the problem is changed. Consider a model that explains quantity demanded as a function of price of a good. If the price increases by one dollar, how much will the quantities demanded decrease? An attempt to write this question in the form of a model results in a stalemate for a typical economist wishing to stick to vocabulary that has economic meaning. In equation (1.2) below, the problematic value is designated by "?" The value that replaces "?" answers the question "if the price increases by $1, (how much) will the quantity demanded decrease." The "(how much)" in the parenthesis does not have a defined economic name, thus, for the time being it is represented by a question mark.

$$\text{Quantity demanded} =$$
$$\text{demand when the good is free} + (?) \times (\text{price}) \qquad (1.2)$$

The "?" can be replaced by "responsiveness of quantity demanded," or some other unfamiliar and arcane wording. Such arbitrary naming can only cause confusion and should be avoided. A "reasonably good" alternative for the (?), which would be close to the concept of *MPC* in equation (1.1), could be "coefficient of responsiveness of quantity demanded to changes in price." One advantage of this term is the use of the previously defined concept of **coefficient**. While this phrasing still has the shortcomings of the previous naming, it also has the added disadvantage of being long and wordy. Furthermore, an astute student would recall that it resembles the definition of **elasticity**. In fact, had the price and quantity been measured in units of natural logarithm, the question mark could be replaced by **price elasticity**, as demonstrated in equation (1.3).

$$\ln(\text{quantity demanded}) = \text{demand when the good is free} +$$
$$(\text{price elasticity of demand}) \times (\text{price}), \qquad (1.3)$$

where "ln" indicates natural logarithm as is customary. Sometimes equations that involve natural logarithm on both sides of the equation are called log–log, but this is a poor and inappropriate terminology, as is the name double-log equation.

Definition 1.5

Price elasticity of demand is the percentage change in quantity demanded divided by the percentage change in price.

By expressing the price and quantity in natural logarithm, the coefficient of the slope of the price variable becomes the same as the demand elasticity. This is due to properties of the slope of regression line and mathematical properties of the natural logarithm. In Chapter 9, using logarithm we address some modeling and data problems. In equation (1.3) there is no good explanation for intercept, so for simplicity and brevity it can be called by its generic term, namely the intercept. Nevertheless, it is better to think of the model in economics terms as much as possible.

Although writing models in their economics equivalent terms is extremely useful, it can also be a cumbersome process. At times, it is helpful to use symbols instead of words. For example, if we replace consumption with C, income with Y, and marginal propensity to consume with MPC in equation (1.1), as is customary, we obtain the following equation:

$$C = \text{subsistence level of consumption} + (MPC) \times (Y) \qquad (1.4)$$

One might choose to represent subsistence level of consumption with "*SLC*," but the acronym is not customary and thus, it does not help much. A more generic symbol might prove more pragmatic.

Parameters are customarily represented by Greek letters, which make most people apprehensive. Consider the Greek letters as names for parameters, which are generic terms. Equation (1.4) can be written as

$$C = \beta_0 + \beta_1 Y \qquad (1.5)$$

A novice mathematics student might be ill at ease with equation (1.4) or (1.5) because in mathematics it is customary to use the letter Y for the dependent variable, while here it is used to represent the independent variable. Economists customarily use the letter Y for income and are fairly comfortable with it. However, the following format is not only preferred but also more informative:

$$\text{Consumption} = \beta_0 + \beta_1 \times \text{income} \qquad (1.6)$$

This indicates that if income changes by one unit, consumption changes by β_1 units in the direction of the sign of β_1, which according to consumption theory, should be positive. This theoretical expectation of the outcome is the foundation of forming the alternative hypothesis. For more information consult.[1] For example, if β_1 is 0.8, then as income increases by \$100, consumption will increase by \$80. This "expected outcome" can be verified empirically, which makes it a testable hypothesis. In order to test the magnitude of the *MPC*, the slope parameter (β) must be estimated, as will be discussed later. The next step after estimating a parameter is to test the estimated value against theoretical expectation. In this example, it makes sense to test the estimate of the parameter to determine if it is equal to the numeral one, which indicates zero savings and zero borrowing. As it will become clear later, it would also make sense to test the estimated slope against the value of zero.

From a Mathematical Equation to a Regression Model

None of the equations that have been presented thus far are actually **regression models**. They are **mathematical functions** and more specifically, each is an equation of a line. Equations (1.1) and (1.4)–(1.6) are consumption lines, where consumption is a function of income, while equation (1.2) is a demand line or function. Equation (1.3) is a line representing the percentage change in quantity demanded as a function of percentage change in price. Its main parameter is the price elasticity of demand, which is the coefficient of the independent variable percentage change in price.

The reason none of these equations are models is that they are exact mathematical equations, as depicted in Figure 1.1a, and not a simplification of a real phenomenon in life. Things in real life occur with a degree of uncertainty or probability and thus, they are random in nature. Adding a **random component** to these equations converts them into a regression model. The random component is called **error term**, or **random error**, for reasons that will be explained shortly. The customary symbol is the Greek letter **epsilon** (ε), but (U) and (V) are also common. In Figure 1.1b, the vertical distances between the actual observations and the regression model are the error terms.

$$C = \text{subsistence level of consumption} + (MPC) \times (Y) + \varepsilon \quad (1.7)$$

$$\text{Consumption} = \beta_0 + \beta_1 \text{ income} + \varepsilon \quad (1.8)$$

$$C = \beta_0 + \beta_1 Y + \varepsilon \quad (1.9)$$

The *above* three equations (1.7)–(1.9) are **regression models** and express exactly the same thing. They are models that state, on the average, consumption depends on income in a linear fashion. These are all the same as claiming that income explains average consumption. Note that the use of the term average refers to average outcome for a dependent variable, which because of random error is probabilistic in nature and has an average. It is different than the concept of average consumption, which is consumption divided by income.

Soon you will learn that having a model is not sufficient; a model must be useful, which is a concept that needs to be defined and clarified. For sake of completeness, the dependent variable (C) represents consumption. For slope, we use the acronym MPC. The independent variable (Y) represents income. Epsilon (ε) is the **error** term; β_0 (beta zero) is the intercept, which represents the subsistence level, and β_1 (beta one) is the slope, which in this case represents the MPC.

Students and scholars should develop the habit of following the same procedure for regression models as it is customary in the profession. The dependent variable, what is being explained, appears on the left-hand side of the equal sign. Examples from the above models include consumption, quantity demanded, and percentage change in quantity demanded. The term that is not related to the independent variable, the intercept, appears as the first term on the right-hand side of the equal sign. It represents the value of the dependent variable in the case where the independent variable fails to be significant, which is reflected by a zero value for its coefficient.

The **independent** variable and its **coefficient** are next on the right-hand side of the equation. In the three examples above, there is one independent variable in each model. The independent variable for the consumption model is income, while for the quantity demanded model it is price. Finally, for the model estimating elasticity, the independent variable is the percentage change in price. If there were more than one independent variable, as will be the case soon, the variables follow the same pattern one after the other but not necessarily in any particular order. In fact, the order in which independent variables are listed in a model has no impact on the final output. The coefficient of the independent variable is also called **slope** of the line; however, it only makes sense if there is only one independent variable, as has been the case with the examples so far.

Customarily, the last term is the **error term or ε**, which plays a very important role in a regression model. It converts a mathematical function into a regression model that can be estimated using statistics. For a regression analysis to be valid, the **error** term must comply with certain requirements, which are customarily called **assumptions**. The assumptions are placed in the appendix because of the theoretical nature of the discussion.

The Meaning of Regression

As noted earlier, equations (1.1) and (1.4)–(1.6) state the same thing, while models (1.7), (1.8), or (1.9) are exactly identical. We choose equation (1.6) and model (1.8) for comparison. The difference between an equation like (1.6) and model (1.8) seems to be that model (1.8) has one extra term, namely, the (ε), which we learned is called the error term. However, there are a number of major differences between the two equations. Some are simplistic, such as the fact that equation (1.6) is a mathematical function, while equation (1.8) is a regression model. The other differences need more explanation, which should clarify the difference between an equation and a model. A mathematical function represents an exact relationship with exactly the same outcome each time it is performed. However, a model is a representation or simplification of reality and includes a random error term to indicate that the outcome is stochastic rather than deterministic. The term **stochastic** means that a model is probabilistic in nature; therefore, every time a new sample is obtained and the regression model is estimated, the results are slightly different, reflecting the random nature of the model.

In equation (1.6), the parameters β_0 and β_1 are known. In contrast, in model (1.8) they are unknown and must be estimated. The customary use of equation (1.6) is to find the value of consumption with knowledge of known parameters β_0 and β_1 and a given value of income. The fact that β_0 and β_1 are known means anyone who chooses to insert a given value of the independent variable income in the equation would always get the same answer. No real data is necessary. If one chooses to use real data such as per capita income for a country for years 1973–2010, it is possible to obtain one value for consumption for each year. On the other hand, in model (1.8) the parameters β_0 and β_1 are unknown, which means it is impossible to obtain a value for consumption even with a known value for income until parameters β_0 and β_1 are estimated using regression analysis. In using model (1.8) the data for consumption and income are available. They are historical values that have been observed and cannot be changed or replaced arbitrarily. Using these observed values the objective is to estimate the unknown parameters to obtain a line that best fits the data. The study of regression analysis deals with methods for obtaining estimates for β_0 and β_1 that meet certain criteria deemed desirable and also to determine if there is a set of estimates that is **best**; a concept that must be defined clearly and precisely and will be covered in Appendix A. Customarily, estimated parameters are represented by Greek letters with a "^," called a "hat" symbol, as $\widehat{\beta}_0$ and $\widehat{\beta}_1$. These are pronounced **beta-hat-sub-zero** and **beta-hat-sub-one**, respectively.

A model represents a **claim** about a real-life phenomenon. For example, model (1.8) claims that there is a cause and effect relationship between income and consumption, that is, as income increases consumption increases. One cannot include *vice versa* at the end of last sentence, because based on economic theory it is not true. In economics, income determines consumption while consumption does not determine income, at least not in an introductory discussion of the subject. The theory that states income determines consumption belongs to economics not statistics. The fact that in macroeconomics, consumption also depends on income, via a different mechanism, is addressed later in a much more sophisticated analysis in more advanced economic courses. A model, as a simplification of reality, is proposed to explain the causal relationship between income and consumption. Regression analysis, as a statistical tool, is used to provide a theory that determines if there is sufficient evidence in real life to support the claim presented in economics. The theories that justify inference based on evidence belong to statistics not economics.

Therefore, every research model involves two different types of theories, one from the discipline in which the research is conducted and the other from statistics. The starting point for every research is the theoretical foundations of the discipline, which for us is economics. The estimation and inference of the research are governed by theories in statistics. The first set of theories originates in economics, which provides the foundation for raising the research question and establishing the claim(s) of the study. For research in other fields, the relevant subject provides the appropriate theory for this purpose. Statistical theories govern the procedures and assure that outcomes have desirable properties and can be generalized. Some of the desirable properties will be explained and verified in this manuscript. Lack of appropriate theories from either the field of economics or statistics invalidates the research outcome.

A consumption model like equation (1.8) is used to determine whether there is empirical evidence to refute economic theory. Note that economic theory does not make any assumption that parameters β_0 and β_1 are known. Although it places restrictions on them, such as β_1 must be a value between 0 and 1, when β_1 represents MPC. Any number outside the range 0 and 1 violates one or more economic rules or principles. A slope greater than 1 means that a one unit increase in income would increase consumption by more than 1 (for example, if β_1 is 1.2, then a $1.00 increase in income would increase consumption by $1.20), which at least in this simplest of consumption models is impossible. Also, a negative MPC makes no economic sense. Theoretical properties of the coefficient can also be tested statistically, as will be seen in Chapter 6.

In order to test any theory using a model there must be sufficient data. Because parameters of the proposed model (β_0 and β_1) are unknown, a statistical method known as regression analysis is necessary. Regression analysis is also called the **method of least squares**. The simplest regression analysis uses a model that has only **one independent variable**, such as income, which means it has **two parameters**, β_0 and β_0. These parameters are also known as intercept and slope, respectively. This simple regression analysis requires one set of data, customarily arranged in two columns, one for the independent variable and another one for the dependent variable, which in this case are income and consumption, respectively. **Estimated parameters** depend on a particular observed set of data and are shown as $\widehat{\beta}_0$ and $\widehat{\beta}_1$.

CHAPTER 2

The Method of Least Squares

The Logic Behind Regression Procedure

The method of least squares, or the regression analysis for a model with one variable, is sometimes called the *ordinary least squares (OLS)*. In this method a line, similar to equation (1.1) in Chapter 1, is claimed to represent real-life data such as the income and consumption data for the US for years 1990–2010, which are plotted in Figure 2.1. Each dot on the graph represents a pair of income–consumption data for one year. The data are not sorted because doing so would not create any difference in the analysis. An attempt to have a line graph where data points for consecutive years are connected will usually present a meaningless graph. Therefore, the customary way of presenting the data is a *scatter plot*, as in Figure 2.1 where income is presented on the horizontal axis and consumption is presented on the vertical axis.

Because pairs of income–consumption data for years 1990–2010 do not line up perfectly, some compromise must be made if one wishes to use a single straight line. The compromise must be logical. In fact, you have been using such a logical line for a long time, as will become clear shortly.

It is not unusual to report typical consumption. One way of defining *typical* is to use *average* consumption. Average consumption, which is actually the sum of consumption values divided by the number of observations, is not a function of income. That means regardless of the level of income, the average consumption does not change. The graph for average consumption on an income–consumption coordinate is a flat line parallel to income axis at the level where consumption is equal to the mean value

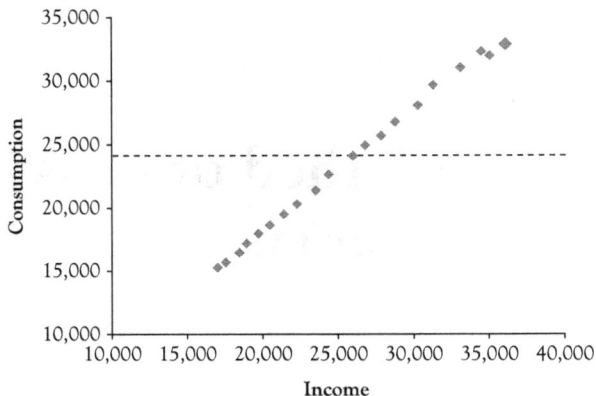

Figure 2.1. Scatter plot of income–consumption for the US (years 1990–2010).

of consumption. Because the average of consumption does not depend on income, all the income–consumption pairs line up at the level of the average on the consumption axis, while the line connecting them is horizontal and parallel to the income axis. The flat dashed line in Figure 2.1 represents the *average income*, which is $24,137. This line represents a model, which provides an unbiased estimate of the average for consumption. The line is the best estimate when no other contributing factor such as income is considered. Model (1.8) from Chapter 1 is reproduced below as equation (2.1).

$$\text{Consumption} = \beta_0 + \beta_1 \text{ Income} + \varepsilon \qquad (2.1)$$

If the marginal propensity to consume $MPC = \beta_1 = 0$, then consumption function becomes a constant, and the graph of a constant on a scatter plot is a flat line. This line, by virtue of representing consumption data, will be in the middle of the scatter plot of consumption–income data points and can serve as a model representing the data.

When income is completely ignored, for example, when $MPC = \beta_1 = 0$, then the parameter of interest is the average consumption. The regression estimate of such a model is presented in equation (2.2)

$$\widehat{Consumption} = \widehat{\beta}_0 \qquad (2.2)$$

It can be proven that $\widehat{\beta_0}$ from equation (2.2) is actually the estimate of average consumption; the graph of which is the dashed horizontal line in Figure 2.1. As is evident, obtaining an estimate for a model that represents income–consumption is as easy as finding the average of consumption data. Therefore, a possible regression line for model (2.1) is the average of the dependent variable. A question remains unanswered. How good is this flat line in explaining real data? It turns out that the average of the dependent variable is not a "good" regression line. The reason is that many other alternative lines, and specifically, the regression line obtained by the *method of least squares*, provide "better" results. The concept *better* refers to the smallness of the average squared error.

There are several error concepts, which are closely related. We will define some of these concepts first. We then use an appropriate error concept to demonstrate that the regression line obtained from the method of least squares is better than the one obtained from the average of the dependent variable. The first error concept is the deviation of observations from the mean, which is called *individual error*.

Definition 2.1

An individual error is the difference between an observed value and its expected value.

When only consumption is considered and the contribution of income in explaining variations in consumption is ignored, the expected value is represented by the mean of consumption. The differences of observed values from population mean result in 21 individual errors, depicted in Table 2.1, where "I" is income, "C" is consumption, "μ_C" is average consumption, $(C - \mu_C)$ is the difference between consumption and its average, $\hat{C}$, pronounced C hat, is the regression estimate of Y or estimated consumption, and $(C - \hat{C})$ is the deviation of consumption from its regression line. Comparison of the sums of $(C - \mu_C)$ and $(C - \hat{C})$, the deviations of consumption from its average and from its regression on income, is useless, because both columns sum to zero, as is evident at the bottom of the columns labeled $(C - \mu_C)$ and $(C - \hat{C})$, respectively. Note that both the mean of consumption (μ_C) and the regression estimates ($\hat{C}$) are *expected values* of consumption based on two different procedures.

Table 2.1. Deviations from Average and Regression Line

Year	I	C	μ_C	$C - \mu_C$	$\hat{C}$	$(C - \hat{C})$	$(C - \mu_C)^2$	$(C - \hat{C})^2$
1990	17,004	15,331	24,137	−8806	15,454	−123	77,549,770	15,139
1991	17,532	15,699	24,137	−8438	15,954	−256	71,206,616	65,372
1992	18,436	16,491	24,137	−7646	16,811	−320	58,463,424	102,582
1993	18,909	17,226	24,137	−6911	17,260	−34	47,764,730	1135
1994	19,678	18,033	24,137	−6104	17,989	44	37,264,359	1950
1995	20,470	18,708	24,137	−5429	18,739	−31	29,476,661	982
1996	21,355	19,553	24,137	−4585	19,578	−26	21,018,301	652
1997	22,255	20,408	24,137	−3729	20,431	−23	13,904,451	525
1998	23,534	21,432	24,137	−2705	21,643	−212	7,318,393	44,777
1999	24,356	22,707	24,137	−1430	22,423	285	2,044,272	81,068
2000	25,944	24,185	24,137	48	23,928	258	2332	66,352
2001	26,805	25,054	24,137	917	24,744	310	840,750	96,169
2002	27,799	25,819	24,137	1681	25,686	133	2,827,433	17,558

2003	28,805	26,833	24,137	2695	26,640	193	7,265,020	37,184
2004	30,287	28,179	24,137	4042	28,044	135	16,336,868	18,122
2005	31,318	29,719	24,137	5581	29,022	697	31,152,330	485,680
2006	33,157	31,102	24,137	6964	30,765	337	48,503,972	113,461
2007	34,512	32,356	24,137	8219	32,049	307	67,548,602	94,127
2008	36,166	32,922	24,137	8784	33,617	-695	77,165,425	483,546
2009	35,088	32,087	24,137	7950	32,595	-508	63,194,818	258,518
2010	36,051	33,039	24,137	8902	33,508	-469	79,237,909	220,137
Total	549,461	506,880	506,880	0	506,880	0	760,086,436	2,205,035

Sources: Bureau of Economic Analysis, National Income and Product Account Tables: Table 2.3.5—*Personal Consumption Expenditures by Major Type of Product. Bureau of Economic Analysis, GDP and Personal Income: SA1-3 Personal Income Summary.*

Rule 2.1

The sum, and thus the average, of deviations of values from their expected value is always zero. In other words, sum of individual errors is always zero.

This is due to cancelation of negative and positive errors. One way to avoid this outcome is to square the deviations, which are presented in Table 2.1 in columns with headings $(C - \mu_C)^2$ and $(C - \hat{C})^2$, respectively.

Note the following relationships. The total for consumption is the same as the total for the average of consumption repeated 21 times in the column named mean consumption (μ_C) and the column of estimated consumption $(\hat{C})$. The reason for the sum for estimated consumption being equal to the actual sum is that on average there is no error in regression estimates. Note that the *sum of squared deviations of consumption from mean of consumption* $(C - \mu_C)^2 = 760,086,436$ is much larger than the *square of deviation of consumption from regression estimates* $(C - \hat{C})^2 = 2,205,035$. Comparing the results of the sums of squared (SS) values in the last two columns of Table 2.1 is unreasonable because the values are based on different numbers of observations, and therefore, different numbers of *degrees of freedom*. The totals for the last two columns of Table 2.1 are part of the regression output. Regress consumption on income and compare your results with those in Table 2.2. Instructions for performing regression analysis in Excel are provided in Chapter 3.

Explanation of Output

Detailed explanation of the output is provided in Chapter 5. Here we only focus on two values, the *sum of squares of residual* and the *sum of squares total*. It is alright to refer to these as *residual sum of squares* and *total sum of squares*, respectively. They are in rows two and three of column "SS" in the ANOVA section in Table 2.2. Note that these two values are exactly the same as the sum of the squares of deviations of consumption from regression line $(C - \hat{C})^2 = 2,205,035$ and the sum of the squares of deviation of consumption from its average headings $(C - \mu_C)^2 = 760,086,436$ in Table 2.1, respectively. It is noteworthy that regression SS values is based on one degree of freedom, residual SS is based on 19 degrees of

Table 2.2. *Output for Regression of Consumption on Income for the Years 1929–2010 in the US*

SUMMARY OUTPUT

Regression of Per Capita Personal Consumption Expenditure on Per Capita Disposalbe Income in Current $

Regression statistics	
Multiple R	0.9985843
R Square	0.99709897
Adjusted R Square	0.99694628
Standard Error	340.667694
Observations	21

ANOVA

	df	SS	MS	F	Significance F
Regression	1	757881401	757881401	6530.39343	1.41798E–25
Residual	19	2205035	116054		
Total	20	760086436			

	Coefficients	Standard error	t stat	p-value	Lower 95%	Upper 95%	Lower 95.0%	Upper 95.0%
Intercept	–663.52392	315.7729644	–2.1012689	0.04919061	–1324.444326	–2.6035057	–1324.44433	–2.60350566
Income	0.94786316	0.011729405	80.8108497	1.418E–25	0.923313234	0.97241309	0.923313234	0.972413087

Abbreviation: ANOVA, analysis of variance.

freedom, and total sum of squares is based on 20 degrees of freedom. This makes the comparison between the two values meaningless. To make these comparable, *SS* values are divided by their corresponding degrees of freedom to obtain *mean squared* (*MS*) values. As you see there is no *MS* value reported for the *total*.

Sum of squared total represents total variation in the dependent variable. In your earlier statistics course,[1] you dealt with this value as the numerator of the *sample variance*. The sample variance consists of squared values of individual errors divided by degrees of freedom. It shows the amount of variation in the dependent variable that cannot be explained by the mean of the dependent variable. To verify this outcome, calculate the variance of consumption for the data in Table 2.1. The command in Excel is

$$Var.s(c2:c22) = 38,004,321.78$$

Dividing Total *SS* by degrees of freedom = 760,086,436/20 = 38,004,321.78
In older versions of Excel the command was

$$Var(c2:c22) = 38,004,321.78$$

Both numbers are rounded up to two decimal places. Part of the total sum of squared, or the previously unexplainable variation in the dependent variable, the value 760,086,436, can now be explained by the regression model. This amount, represented by regression *SS* is 757,881,401, which is displayed under *SS* column on the row for "regression." This value is called *sum of squares regression* or *regression sum of squares*. As stated earlier, although this is a substantial difference it is misleading due to the fact that different numbers of values are used to get these sums. The sum for regression has one degree of freedom, while the sum for the total has 20 degrees of freedom. Other things equal, the sum of more numbers would be greater than the sum of fewer numbers. To determine whether the portion explained by regression is statistically significant, the regression sum of squares and the residual sum of squares are divided by their corresponding degrees of freedom. The customary comparison is between the portion *explained* and the portion that is *unexplained*, as will be discussed in detail in Chapter 6.

Residual sum of squares is the amount of variation in the dependent variable that is *unexplained* by either the mean of dependent variable or regression line. Here too, the appropriate value is the *average* value instead of the sum. To obtain average value of *regression sum of squares* and *residual sum of squares*, divide them by their respective degrees of freedom, which are 1 and 19 for this example. The results are called *regression mean square* (757,881,401) and *residual mean square* (116,054), respectively. These terms can also be called *mean square regression* and *mean square residual*, respectively. It is not unusual to refer to *residual* as *error*, in the above terms.

Verify that *MS* for regression and residual is obtained by dividing their respective *SS* by the corresponding degrees of freedom for data in Table 2.2. We will demonstrate this in an example in Chapter 3. The ratio of mean-squared regression to mean-squared residual provides the value of the *F statistics*. Dividing $757,881,401/116,054 = 6,530$, which is provided in Table 2.2 in the regression output.

The above exercise demonstrates that regression analysis can reduce the amount of unexplained variations in the dependent variable. *F statistics* is the measure that verifies whether the explained portion of the dependent variable exceeds the unexplained portion. *F* statistics is discussed in more detail in Chapter 4 and more formally in Chapter 5.[1] The above practice demonstrates that regression line is better than the average of the dependent variable in explaining variation of values of dependent variable. However, there are numerous other statistics that can be used to explain this variation. We need to establish which one produces the *smallest average error*. The idea of finding the smallest possible average of errors is a novel one. Unfortunately, it is useless if the deviations from mean are used because, as we showed earlier, the sums of deviations of observations from their mean or deviations from regression line are both zero. There is no smaller average error than zero error. The problem is not in the novelty of using the method; it is in the fact that errors cancel each other out. One remedy is to square the errors to avoid such cancellation, which was used in the above demonstrations. The sum of squared errors (*SSE*) is never zero unless all the errors are identical and equal to zero, which can happen if only if all the observations have the exact same values—another boring and unrealistic case. There are mathematical methods to find estimates of the parameters of model, namely β_0 and β_1, such that the estimated equation would have the *least squared errors*, which is the origin of the name of the *OLS* as an alternative to the name *regression*.

Minimizing the Squared (Individual) Errors

This section explains the concept and the method of minimizing squared values of individual errors. As explained above, individual errors are not identical, reflecting the random nature of real-life data. There are as many individual errors in any given dataset as the number of pairs or rows of data. Because in statistics we prefer to capture the nature and the essence of data in as few parameters as possible, it is better to use the average of individual errors. As the *sum of all individual errors is always zero*, their average is also zero. Squaring each individual error resolves this problem. *SSE* is always non-negative. The trivial case of zero sum of squared errors is excluded. Consequently, the more the number of observations the greater is the *SSE*, with other things being equal. This is the reason for dividing sum of squared residuals by its degrees of freedom, which yields average squared error. A more important factor affecting the size of *SSE* is the choice of the estimator. As we saw earlier in this chapter, you have been using the mean of the dependent variable, which is consumption in the above example, as a possible regression line. We also established that having information on variable(s) that influence the dependent variable, in this example the income, will reduce the amount of unexplained portion of the dependent variable, enabling the researcher to provide a better estimate for the dependent variable. The term *better* means *smaller variance*, or less error. Better estimates result in better forecasting.

When only consumption data is considered, sample consumption mean provides the *best* regression line; nevertheless, it ignores the contribution of factors that economics theory offers, such as income. Note that in this context the line is a horizontal line with zero slope. Taking advantage of the extra information provided by income and the use of the regression analysis improves the estimate. The regression line explains part of the previously unexplainable variation in the dependent variable. Because part of the previous error is explained by regression, it provides a better estimate than the average of the dependent variable. Once again, the adjective *better* is used to indicate *smaller variance*.

The regression line intersects with the line representing the average of consumption at exactly the average points of both income and consumption. The intersection of the two lines is the center of data. The center of data is actually its center of gravity. If each observation, pairs of income—consumption points, is given the same weight and is placed on a plane, the center of data is the point where the plane is balanced on a pin.

The simplest way to obtain estimates of parameters of model, namely β_0 and β_1, is to find partial derivative of *SSE*, set them equal to zero, and then solve the two resulting equations to obtain the values of β_0 and β_1 that minimizes *SSE*. This requires knowledge of calculus. Some of the books written for students majoring in statistics, mathematics, or closely related fields, provide derivation of the estimates of parameters β_0 and β_1 and mostly place them in their appendix.[2] We will not concern ourselves with such derivations.

The method of least squares provides estimates for parameters β_0 and β_1 that minimize squares of individual errors. Note, however, that in the definition below

$$\text{Individual error} = \text{observed} - \text{expected}$$

The "expected" is no longer the mean or the average of the data; it is the regression estimate of the dependent variable.

The estimated values of dependent variable and parameters are represented by a "hat."

$$\hat{Y} = \widehat{\beta_0} + \widehat{\beta_1} X \qquad (2.3)$$

Thus, now

$$\text{Individual error} = Y - \hat{Y} \qquad (2.4)$$

For the consumption example, these values are shown under column $(C - \mu_c)$. Individual errors are represented by vertical distance between the

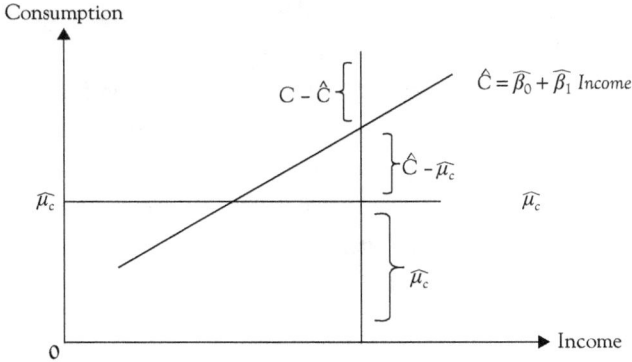

Figure 2.2. Deviation of observations from mean and regression line.

observed value and the regression line in Figure 2.2; only some of the lines are shown to avoid clutter. Note that in some cases the distance between an observation and the regression line exceeds the distance between the same observation and the line representing the mean of dependent variable, which is consumption in our example. Nevertheless, mean of *SSE* obtained from regression line is the smallest average of squared errors including the one based on the average of dependent variable, which is also depicted in Figure 2.2 as $\widehat{\mu_c}$.

Error

It is of vital importance not to mistake any of the concepts of errors discussed here with errors in measurement. *Errors in measurement* refer to incorrectly measuring or recording the values of dependent or independent variables. In some social science disciplines, lack of measurement error is called *validity*.

The notion of error plays an important role in statistics in general, particularly in regression analysis. We have seen individual error, error term in regression model, *SSE*, and mean of squared error (*MSE*). As one might expect there are close relationships between these concepts, some of which have been explained already. This section provides an opportunity to solidify these concepts and provide further insight into the error concept.

Definition 2.2

Error in statistics is what we are unable to explain. It is the difference between an observed value and its expected value.

The observed values are collected data about a phenomenon, such as consumption. The *expected value*, however, depends on the theory we use and the data that are available. For example, in the case of a single variable, the expected value is its mean. Therefore the error is

$$\text{Error} = Y - \mu,$$

where μ, pronounced mu, is population mean. When population mean is not available its estimate, the sample mean, is used.

$$\widehat{\text{Error}} = Y - \hat{\mu} \qquad (2.5)$$

An observant student will notice that the above two formulas were called *individual error* earlier to focus on the fact that for each row of data there is one such error.

When a regression model is proposed, the claim is that the regression line explains part of what was previously unexplainable. Individual errors can be represented by the vertical distance between an observed point and the flat line that represents the average. The claim of the regression model is that the vertical distance reflecting the error can be partitioned into the segment explainable by the regression line and a smaller segment, on the average, that still cannot be explained, as depicted in Figure 2.2. Representing the estimated regression equation by $\hat{Y}$, the individual error is the same as equation (2.4)

$$\text{Individual error} = Y - \hat{Y}$$

These individual errors are shown as distances between observations and regression line instead of the line for average consumption in Figure 2.2. The previous individual error $(Y - \hat{\mu})$ can no longer be called error because at least part of it is explained by the estimated regression equation. The partition can be demonstrated very easily by the following identity:

$$Y - \hat{\mu} = Y - \hat{\mu} \tag{2.6}$$

Subtract and add the amount $\hat{Y}$ from the right-hand side of the equation. Because the same amount is added and subtracted, the identity remains valid.

$$Y - \hat{\mu} = Y - \hat{\mu} + \hat{Y} - \hat{Y} \tag{2.7}$$

Rearrange the right-hand side to obtain

$$Y - \hat{\mu} = (Y - \hat{Y}) + (\hat{Y} - \hat{\mu})$$

The left-hand side is the deviation of observed values of dependent variable from their mean. The first term on the right-hand side represents the amount that is explained by regression, and the second term is what is left unexplained. It can be proven[2] that squaring these individual errors and summing them up results in the following:

$$SST = SSR + SSE,$$

where SST is sum of squared total, SSR is sum of squared *regression*, and SSE is sum of squared *errors* or sum of squared *residual*. These are the three components in the column designated as SS in Excel output that were discussed and analyzed above in theoretical format and will also be discussed in Chapter 3 using examples.

Dividing SSE by the corresponding degrees of freedom, which is $n - k$, that is, number of (pairs) of observation minus number of parameters (β_0 and β_1, or 2 for the simplest case), results in MSE, which is the same as *variance*. Therefore, variance of a regression model is its MSE. Hence, square root of MSE is the *average error or regression*.

CHAPTER 3

Simple Linear Regression in Excel

Regression in Excel

To complete the exercises in this chapter, you will need a Microsoft Excel add-in called Analysis ToolPak. The add-in, which is included in the software, is not installed automatically. To check whether you have it installed, open the Data tab in Excel. Analysis ToolPak should be at the far right-hand end of the ribbon. In the newest version it appears as Data Analysis. If it is not installed, please follow the instruction provided with your specific version of Excel software. Press the "F1" key to activate the "Help" screen in Excel. Type Analysis Tool Pack in the search window and follow the instructions. There is no need to type the quotation marks. Print the help instructions to facilitate following the instructions without having to move between windows.

Once installed, you can access Analysis ToolPak from the Data tab in Excel. If you have not installed any other add-ins for your software, Analysis ToolPak will be the last option at the right-hand corner of the Data tab, in the 2007 and 2010 versions of Excel.

Choosing the Data Analysis option (shown at the top right in Figure 3.1) opens a pull-down menu labeled Data Analysis with 19 analytical tools (Figure 3.1). In order to reach the regression option, either scroll down the list or press the letter "r" on your keyboard three times.

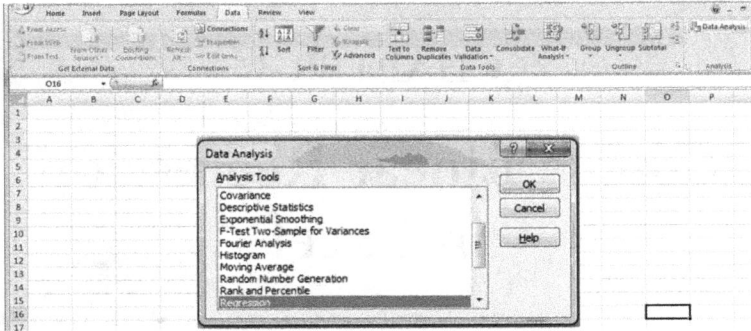

Figure 3.1. Data analysis pull-down menu.

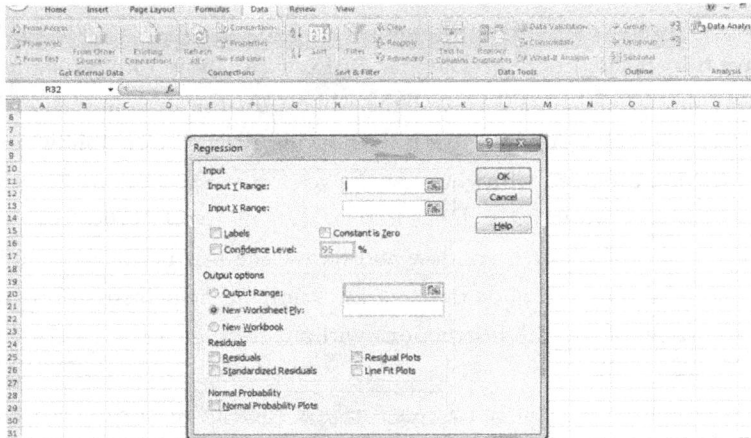

Figure 3.2. Regression window.

Double clicking on the regression option will open the regression window (Figure 3.2).

Although pressing "OK" will also open the pull-down menu, it is better to use the mouse instead. Get into the habit of not pressing the *Enter key* until you are completely done with all the options on the drop-down menu and ready to perform regression analysis. Even then, selecting "OK" instead of pressing the Enter key will help to reinforce the habit. If you inadvertently press the Enter key before

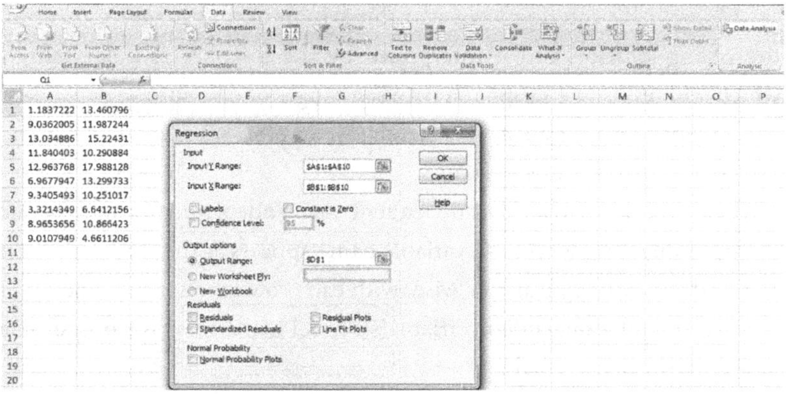

Figure 3.3. Regression data range.

all the options are selected it is nothing to be concerned about. The result is a minor setback because you will get an error message and just have to start over.

The minimum requirement to perform a regression analysis is to insert the Y range and the X range, which is Excel's terminology for the dependent and independent variables, respectively. You are free to provide the location of the data either by pointing to a range, pinning one end by holding the shift key and then dragging the highlighter to the end of the range by moving the mouse or you may actually type in the range by providing the beginning cell address, inserting a "colon," and typing the location, by specifying the cell address, of the end of the range. An example is shown in Figure 3.3.

When data contains variable names, which is recommended, it is important to select "Labels" on the Regression window by clicking on the selection box to its left. Otherwise, you will get an error message and have to start over.

Choosing an output range will become a source of difficulty, sooner or later, because selecting "Output Range" by choosing the "Radio Button" does not place the cursor in the input box on the right-hand side, as expected. Instead, the cursor would remain in the last input box location, which would be either the Y range or X range, depending on which one was modified last.

Caution: By entering what you think is an output range, ends up in the wrong location and is interpreted by the software as a data range, which will result in a mismatch between ranges of data for the dependent and independent variables. Consequently, an error message pops up and you will have to start over again.

Preferably, the data for dependent and independent variables are arranged in columns where a variable name appears at the top, in the first row and the cases appear row wise, with each row representing one set of dependent and independent variable(s). Excel will display the output in a new worksheet by default.

Excel does not provide information about the dependent variable in the output, only the name(s) of the independent variable(s) is/are reported. If you run different regressions with different dependent variables and place them on the same worksheet with different ranges, it will become confusing very quickly. Furthermore, if you choose the "Output Range" you must provide a different output range each time, to make sure that the new output does not delete the previous result by overwriting the same range. If you forget to provide a new range, Excel will remind you and will let you choose to delete the previous output or to keep it by providing a new output range. It is recommended that the name of the dependent variable be stated in an empty cell for later reference.

The other advantage of having the output in a new worksheet is that the new worksheet can be labeled appropriately, which will prove beneficial in the long run. You can provide additional information in an empty cell by indicating the name of the dependent variable and other useful information about the procedure, data, project, etc.

An Example

Performing a regression analysis without a theory is a practice in futility. To demonstrate that it is possible to perform a regression without a theory, two columns of random numbers will be generated where one column, arbitrarily called the dependent variable, is regressed on the other column, called the independent variable.

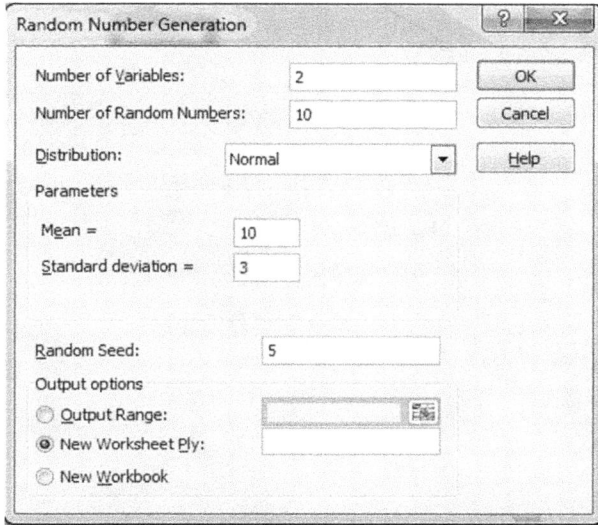

Figure 3.4. Random number generator window in Excel.

Generate Random Numbers

Let us generate two columns of data using a (pseudo) random number generator provided by Excel. The series of commands are as follows:

Data → Data Analysis → Random Number Generation → OK

Enter the appropriate values as indicated below in Figure 3.4. Instead of typing "Normal" in the distribution window, use the drop-down menu and choose "Normal."

The results will be displayed in a different worksheet. There is no need to indicate a *seed* value. This is provided to make sure you get the same data as the one presented here. Computer-generated random numbers are not truly random, because they are generated by an algorithm. The actual outcome depends on the internal clock of the computer, which actually provides the starting value or a *seed* value. Specifying the seed value assures that everyone gets the same set of numbers, regardless how often the command is invoked.

A	B
9.036201	11.98724
13.03489	15.22431
11.8404	10.29088
12.96377	17.98813
6.967795	13.29973
9.340549	10.25102
3.321435	6.641216
8.965366	10.86642
9.010795	4.661121

Figure 3.5. Two columns of random numbers.

For this example, it is helpful for the user to use the same seed number to generate exactly the same data because it allows you to duplicate input to obtain the same output as in Figure 3.5. Random numbers, generated by the previous command using seed five (5), are shown below.

Regression and its Output

The steps for inputting data and performing regression are given below. An example of this process, including data input, is shown in Figure 3.3.

Data → Data Analysis → Regression → OK

In the Regression dialog box, enter the appropriate values as indicated below

Input *Y* Range	A1:A10
Input *X* Range	B1:B10

Detailed explanations of these results are covered in Chapter 2. In this case, the semantics and their representations are discussed.

Table 3.1. Regression of Random Numbers in Column A on Random Numbers in Column B

SUMMARY OUTPUT

Regression statistics

Multiple R	0.332170282
R Square	0.110337096
Adjusted R Square	−0.00087077
Standard Error	3.88569026
Observations	10

Column A is the dependent variable
all data are generated by N(10,9) seed 5

ANOVA

	df	SS	MS	F	Significance F
Regression	1	14.98037077	14.98037077	0.992169919	0.34832699
Residual	8	120.7887518	15.09859397		
Total	9	135.7691226			

	Coefficients	Standard error	t stat	p-value	Lower 95%	Upper 95%	Lower 95.0%	Upper 95.0%
Intercept	4.775695465	3.999175521	1.194170008	0.266615181	−4.46419816	13.99781075	−4.46419816	13.99781075
X Variable 1	0.330580596	0.331882483	0.996077266	0.348382699	−0.434741781	1.095902973	−0.434741781	1.095902973

You may also find that some of the regression results displayed in the summary output are apparent. The output starts with "SUMMARY OUTPUT," which includes "Multiple R" and is equal to the square of correlation coefficient when there is only one independent variable and one dependent variable, as in this case. Go to the worksheet where the two columns of random numbers are located and type the following in any blank cell:

$$= \text{CORREL(A1:A10,B1:B10)}$$

Press the Enter key to obtain the same value as the number in front of the "Multiple R" in the regression output. It is not clear why it is reported as it is not used in any other parts of regression analysis. Squaring the value of "Multiple R" in the output produces the value of R square, when there is only one independent variable. This is verified by observing the value in the next row of output.

The other name for *R Squared* is *coefficient of determination*, which is not used in Excel. Coefficient of determination shows the percentage of variation in the dependent variable that is explained by the regression line, over and above the average of the dependent variable.

The *Adjusted R Squared* provides a correction to R square based on the number of independent variables in the model. Adjusted R^2 is a better measure when different models with different numbers of independent variables are used to estimate the same dependent variable and one model is not *nested* in the other model. A model is said to be nested in another model when all the variables that are in the first model are also available in the second model, while the second model has at least one variable that is not in the first model.

Standard Error is the *average error* of a regression model. In Excel, it is the square root of the number in the cell representing the intersection of the column named *MS*, which stands for *Mean Squares* and the row named "Residual." Verify to make sure.

The entry, *Observations*, reports the number of rows of data, which in this case is 10. The next section presents the table of *Analysis of Variance (ANOVA)*. The table consists of three entries: *Total*, which represents total variations; *Regression*, which represents the portion of variation explained by regression model, and *Residual*, which is the proportion of variation in the dependent variable not explained by regression model.

Each row contains values under different headings. The second column, *df*, stands for *degrees of freedom*. The first two rows of this column always add up to the value of the last row. In this case, the degrees of freedom are 1, 8, and 9 for regression, residual, and total, respectively. The next column, *SS* represents sum of squares. The entries for the first two rows of this column also add up to the value in the third row. The columns *MS* stands for mean square. Each of the values in this column can be obtained by dividing the corresponding value in column "*SS*" by those in column "*df.*"

In the case of the regression model with one independent variable, as in this example, the entries for the first row under columns "*SS*" and "*MS*" are always the same, because a regression with one independent variable has two parameters and the degree of freedom is one less than the number of parameters.

One important value is *coefficient of determination*, which is presented under column *MS* for row "Residual." It is actually the *variance* of regression model, although this is not a commonly used term in this context. The square root of this number, which is also reported under *standard error* on the top left-hand corner and discussed earlier, is the *average error* of regression. The customary name for this number in the context of regression analysis, however, is *mean square error*.

The third section of the output provides information about *coefficients* or the estimates of the parameters of the model. In this case, two parameters are reported. They are represented by the generic terms *intercept*, and *X Variable 1* because we did not provide any names for the dependent or the independent variables. Had we done so, the name of the independent variable would have replaced "*X* Variable 1." It is recommended to choose representative names for the variable names for clarifications. Because Excel does not list the dependent variable you should add it somewhere in the output sheet, for future reference. I always place mine in cell D3 in case I copy the output and do not want to include the heading for the first block of output which reads "SUMMARY OUTPUT." The headings for the columns are mostly self-explanatory. For example, coefficients and standard error are well known to the students of statistics. The column "*t* stat" provides the calculated t statistics for each coefficient. The column "p value," represents the probability of type I error for inference about each coefficient. The concept is explained in detail in Chapter 4. Finally, the "lower" and the "upper" 95% refer to the corresponding values for a 95% confidence interval for each of the parameters, intercept, and slope, respectively.

Real Data Example of Simple Linear Regression

In macroeconomics, consumption is presented as a function of income.

$$\text{Consumption} = f(\text{income})$$

This can be presented as a regression model in the following way:

$$\text{Consumption} = \beta_0 + \beta_1 \text{ income} + \varepsilon$$

The variable names of consumption and income are self-explanatory. Epsilon "ε" is the random error, subject to the requirements listed in Appendix. The betas β_0 and β_1 are parameters of the model and are estimated to provide a numerical link between independent variable and the expected value of consumption. There are numerous measures of income.

For this example, we use per capita disposable income:
Annual data from 1929 to 2010 on personal consumption expenditures and population are available from the Bureau of Economic Analysis. We accessed the data at the following site on November 28, 2011.

http://www.bea.gov/iTable/iTable.cfm?ReqID=9&step=1

Copy and paste the above link on your browser. Once the National Income and Product Account Tables are populated, click on *Section 2— Personal Income and Outlays*. From the list of tables, select *Table 3.3.5. Personal Consumption Expenditures by Major Type of Product (A) (Q)*. To change the data range to start from 1929, click on the "Options" icon, which is located on the top right-hand corner above the table. Once the "Options" icon opens a window, change the Series to "Annual" and change the First Year to "1929" and click on "Update."

Line number "1" on the table represents the annual personal consumption expenditure. You may wish to select the "Download" icon, which is in the same box as the "Options" icon. Choose the appropriate version of Excel. Highlight the row of labels that contains the years and the first row of data, which pertains to annual personal consumption expenditure. Next, choose copy and proceed by opening a new worksheet; in cell A1 click on the right button on your mouse; then select paste special, and finally check the box "transpose." Transpose will display the data as columns instead of rows.

The per capita personal income from 1929 to 2010 is obtained from the Bureau of Economic Analysis using the following link:

http://www.bea.gov/iTable/iTable.cfm?ReqID=70&step=1

Copy and paste the above link on your browser once the GDP and National Income tables are populated. Click on the *Annual State Personal Income and Employment*, and then click on the Personal income and population (SA1-3). Select the *SA1-3—Personal Income Summary* and click "Next Step." Select "United States" for the Area and "All Years" for Year, then from the drop-down list for Statistics, select Per Capita Personal Income (dollars) and click "Next Step."

You may wish to select the "Download" icon, which is next to the "Options" icon in order to generate an Excel file. Highlight the row of labels and the first row of data, which pertains to annual personal consumption expenditure. Next, repeat the previous process by choosing copy; open a new worksheet; in cell E1 click on the right button on the mouse; choose CVS and press OK. Highlight the two new rows and choose the copy command. Place the cursor in cell C2 and again click on the right button of the mouse; choose paste special and finally check the box against "transpose" to get the data to display in columns instead of rows. Now the dates and the data match. You can delete the data from row 1 and row 2 starting in column "E" as they are no longer needed. You should also delete row 1 so the labels appear there. Be sure to save your work.

	A	B	C
1	Year	Personal Consumption Expenditure	Per Capita Personal Income
2	1929	635	683
3	1930	569	605

Enter PE in cell B1 and PI in C1. The best practice is to keep the original download file as it is and save the modified data in a separate spreadsheet or just another worksheet. Label each worksheet accordingly for reference. Make sure income is in column "A" and consumption is in column "B." This assures that following the steps below results in the same output as provided below. To perform a regression, execute the following functions:

$$Data \rightarrow Data\ Analysis \rightarrow Regression \rightarrow OK$$

In the regression dialog box, enter the appropriate values as indicated below:

Input Y Range	B1:B83
Input X Range	A1:A83

Check the box on the left-hand side where it reads "Label" to let the software know the first row is a text containing variable names and not data. Now select OK. The following result in Table 3.2 is provided in a new worksheet. Note that information has been added to the upper right corner of the output for future reference.

Some of the columns are widened to allow their headings to be displayed properly and completely. All output numbers provided by the software depend on the combined 82 pairs of data obtained from the above link. So if you get any number correct, the rest of the results will be correct as well. The output for Multiple R is 0.994,569,105. If your results did not produce this number, then you will need to rerun the regression.

Previously, certain outcomes were linked to others. Now is a chance to verify those relationships and to reinforce our understanding of how those are connected.

Table 3.2. Regression of Consumption on Income

SUMMARY OUTPUT

Regression statistics	
Multiple R	0.994569105
R Square	0.989167705
Adjusted R Square	0.989032302
Standard Error	316.5346259
Observations	82

Regression of Per Capita Personal Disposable Income Current Dollar on Per Capita Pesonal Consumption Expenditure

ANOVA

	df	SS	MS	F	Significance F
Regression	1	731950821.4	731950821.4	7305.32351	2.1886E-80
Residual	80	8015533.552	100194.1694		
Total	81	739966354.9			

	Coefficients	Standard error	t stat	p-value	Lower 95%	Upper 95%	Lower 95.0%	Upper 95.0%
Intercept	-310.002041	46.57998884	-6.65526237	3.2344E-09	-402.699171	-217.30491	-402.699171	-217.30491
Income	0.241953468	0.002830819	85.47118527	2.1886E-80	0.23631996	0.24758698	0.236319959	0.24758698

$$R \text{ Square} = 0.989,167,705 = (0.994,569,105)^2 = \text{Multiple } R$$

$$(\text{Standard Error})^2 = (316.534625)^2 = 100,194.16$$

Regression degree of freedom + residual degree of freedom
= total degree of freedom

$$1 + 80 = 81$$

SS regression + SS residual = SS total

$$731,950,821 + 8,015,533 = 739,966,354$$

MS regression = SS regression/(regression df) = (731,950,821/1)

Your system might have reported the MS regression in scientific nota-tion such as 7.32E+08. This equates to 732,000,000, which is obtained after moving the decimal place to the right by nine places. If you prefer to see the numbers in the customary format expand the column where MS number is reported and you will see the actual number. Unfortunately, you may have to re-format the cell as well.

MS residual = (SS residual)/(SS df) = 8,015,533/80 = 100,194.16

F = (MS regression/MS residual) = 731,950,821/100,194.16 = 7305.32

(Intercept coefficient)/(intercept standard error) = (intercept t stat)

$$(-310.0020/46.57998883) = -6.655262$$

(Disposable income coefficient)/(disposable income standard error)
= (disposable income t stat) = 0.241953/0.002830 = 85.471185

CHAPTER 4

Multiple Regression

The General Model

Economics is complex, and a model with one independent variable is usually not able to explain an economic phenomenon. The extension from simple to multiple regression is easy. The additional exogenous variables are included in the mode linearly. This literally means that independent variables, multiplied by their respective coefficients, are included in additive format in the model.

$$Y = \beta_0 + \beta_1 X_1 + \beta_2 X_2 + \cdots + \beta_K X_K + \varepsilon \qquad (4.1)$$

The subscript K indicates that there are K-many independent variables. K is an integer, which means it is a whole number without a decimal point because it represents the count of independent variables, which is a countable value. The *three dots* represent repetition of the same pattern for the elements in the model where only the subscripts change to represent different independent variables. The model has K independent variable but $K + 1$ *parameters*; one for each of the K slopes and an additional parameter for the intercept (β_0). Independent variables are collected or obtained and are not determined by the model; therefore, the name *independent* or *exogenous* variable, while the parameters of the model are estimated.

β_0 is the intercept. It is the value of Y when all slopes are zero

β_1 is the slope of X_1. A one unit change in X_1 changes Y by β_1, other X's held constant.

β_2 is the slope of X_2. A one unit change in X_2 changes Y by β_2, other X's held constant.

β_k is the slope of X_k. A one unit change in X_k changes Y by β_k, other Xs held constant.

ε is the error term.

The multiple regression model and its error term are subject to the same requirements as of simple regression. In addition, it is assumed that independent variables are not correlated with each other or the error term.

A multiple regression model should not just consist of an arbitrary list of variables. The variables must be included in the model to represent a pertinent theory. Careless inclusion of variables would result in misleading, meaningless, and erroneous conclusions. Many of these problems are demonstrated here and elsewhere in this book.

The term regression line applies only to the case of simple regression, where there is only one independent variable. When there are two or more independent variables, the more appropriate term is regression model.

Interpretation of Coefficients

In the estimated equation, the dependent variable (Y) and parameters (β) of the model are replaced by their estimated values that are represented by the same letters with a *hat* ($\wedge$) symbol over them. The error term (ε) vanishes because the estimated equation represents the *expected* outcome and the expected value or the *mean of errors is zero*.

$$\hat{Y} = \widehat{\beta_0} + \widehat{\beta_1} X_1 + \widehat{\beta_2} X_2 + \cdots + \widehat{\beta_K} X_K \qquad (4.2)$$

$\widehat{\beta_0}$ represents the estimated value of the dependent variable if all the slopes (coefficients) were zero. It can be shown that in such cases $\hat{Y}$ is equal to the average of Y values, which means that the expected value of $\widehat{\beta_0}$ is also equal to the average of Y values when all slopes are zero. This means that the regression model is not explaining anything beyond the average or the expected value of the dependent variable. Each slope represents the effect of a unit change in the corresponding exogenous variable (X) on the endogenous variable (Y), on the average, while keeping all other variables constant. It is important to realize that the above statement is correct assuming all the other variables are already in the model and the variable of interest is the last variable that is added. Therefore, the contribution of a particular slope is based on the contribution of all the other slopes as well. This is the reason for referring to a change in the exogenous variable (Y) as a result of a unit change in a particular X as *conditional expected value*. The correct way of stating the impact is "given that all the other variables are already in the model and are kept constant, the conditional expected value of Y for a one unit change in a particular X is equal to the estimated slope of the variable X."

An Example Using Consumption of Beer

The following data in Table 4.1 provide information on the quantity of beer consumed (Q_{Beer}), price of beer (P_{Beer}), quantity of wine consumed

Table 4.1. Beer Consumption and Price for the Years 1995–2010

Year	Q_{Beer}	P_{Beer}	Q_{wine}	P_{Wine}	Q_{liquor}	I_{Median}	I_{Mean}
1995	22.47	0.81	1.70	4.57	1.35	34,076	44,938
1996	22.51	0.84	1.77	4.93	1.35	35,492	47,123
1997	22.37	0.84	1.79	5.17	1.34	37,005	49,692
1998	22.52	0.86	1.81	5.07	1.33	38,885	51,855
1999	22.74	0.88	1.85	5.24	1.35	40,696	54,737
2000	22.72	0.92	1.90	5.41	1.38	41,990	57,135
2001	22.87	0.96	1.88	5.96	1.38	42,228	58,208
2002	22.98	0.99	2.03	6.23	1.40	42,409	57,852
2003	22.79	1.01	2.08	6.39	1.44	43,318	59,067
2004	22.93	1.07	2.15	6.92	1.48	44,334	60,466
2005	22.70	1.09	2.17	7.77	1.52	46,326	63,344
2006	23.00	1.11	2.22	7.90	1.57	48,201	66,570
2007	23.07	1.12	2.27	8.55	1.59	50,233	67,609
2008	22.98	1.16	2.27	9.45	1.63	50,303	68,424
2009	22.34	1.21	2.27	10.07	1.64	49,777	67,976
2010	22.00	1.23	2.30	9.65	1.66	49,445	67,530

Source: Beer Institute, Brewers Almanac 2001: Per Capital Consumption of Beer By State 1994–2010. US Census Bureau, Table H-6: Regions—All Races by Median and Mean Incom: 1975 to 2010. Bureau of Labor Statistics, Consumer Price Index—Average Price Data: Malt beverages, all types, all sizes, any origin, per 16 oz.

(Q_{Wine}), price of wine (P_{Wine}), quantity of hard liquor consumed (Q_{Liquor}), median income (I_{Median}), and mean income (I_{Mean}), from 1995 to 2010 in the US.

What would be a reasonable model to estimate the demand for beer consumption in the US? Economic theory states that there is an inverse relationship between quantity demanded and price of a good or service, other things equal. The definition clearly provides a causal relationship with the direction of causality from price to the quantity. However, if you consult your economic textbooks for a demand equation, chances are that you will not find any model or even an equation, unless you happen to have an intermediate or advanced microeconomics textbook. The most common form of a demand equation is:

$$Q_d = \beta_0 + \beta_1 P \qquad (4.3)$$

where "Q_d" is the quantity demanded, β_0 is the intercept, β_1 is the slope (coefficient) of the demand curve, and "P" is the price. In this context the demand for beer is:

$$Q_{Beer} = \beta_0 + \beta_1 P_{Beer} \qquad (4.4)$$

However, equations (4.3) and (4.4) are functions and not statistical models. In order to convert them into statistical models, an error term is required:

$$Q_{Beer} \beta_0 + \beta_1 P_{Beer} + \varepsilon \qquad (4.5)$$

Economic theory states that the sign of the coefficient for price of beer should be negative indicating a downward sloping demand curve. This means that the null and the alternative hypotheses for the slope of the price are:

$$H_0: \beta_{P_{Beer}} = 0 \qquad H_1: \beta_{P_{Beer}} > 0$$

The results are depicted in Table 4.2.

Table 4.2. Output of Regression of Q_{Beer} on P_{Beer}

SUMMARY OUTPUT

Regression statistics	
Multiple R	0.062769662
R Square	0.00394003
Adjusted R Square	-0.06720711
Standard Error	0.307600159
Observations	16

ANOVA

	df	SS	MS	F	Significance F
Regression	1	0.005239806	0.005239806	0.05537862	0.817362627
Residual	14	1.324650008	0.094617858		
Total	15	1.329889814			

	Coefficients	Standard error	t stat	p-value	Lower 95%	Upper 95%	Lower 95.0%	Upper 95.0%
Intercept	22.55139171	0.578318785	38.9947418	1.10616E-15	21.31102128	23.79176213	21.31102128	23.79176213
P_{Beer}	0.13084952	0.5697233	0.235326623	0.817362627	-1.0879766	1.356146503	-1.0879766	1.356146503

The estimated equation is:

$$\widehat{Q_{Beer}} = 22.55 + 0.13\, P_{Beer} \qquad\qquad (4.6)$$

In comparison to the theory requirements and implications, the estimated model fails miserably. The goodness of fit test based on the F statistics is not significant, which means the following null hypothesis cannot be rejected.

$$H_0: \text{Model is not good} \qquad H_1: \text{Model is good}$$

The p value for the F statistics is 0.817, which means that if the null hypothesis of "model is not good" is rejected, there is 81.7% chance of a *type I error*. Recall that the type I error means *rejecting* the null hypothesis when it is actually *true*.

Further evidence comes from the coefficient of determination or $R^2 =$ 0.0039. This indicates that only 0.4% of the variation in the quantity of beer can be explained by the price of beer. Finally, the coefficient for the price of beer is positive, while the theory predicts a negative coefficient. One might be tempted to state that the demand for beer has a positive slope and rush to publish a new theory. However, there is no evidence to make such a statement. Because p value (probability of type I error) for the coefficient of the beer price is 0.82, we cannot reject the following hypothesis; otherwise, the probability of committing type I error would be 82%.

$$H_0: \beta_\pi = 0 \qquad H_1 \beta_\pi < 0$$

Therefore, the evidence indicates that the hypothesis that the slope of the price of beer is zero cannot be rejected. In order for an average to be zero, sometimes the estimate must be negative, and sometimes it must be positive. In this particular case, we observed a positive estimate. Nevertheless, the correct slope is zero. At least we are not claiming that the foundation of demand theory is wrong, but having no relationship between the price and quantity is not helping the theory either, not to mention our research.

There are several problems with the above model, one of which is that the requirement of *ceteris paribus* is not met. The way to account for the fact that *other things* do not remain equal and change over time or place is by including them in the regression model. This practice is known as *controlling* for other factors. However, including all the variables that possibly have some effect on beer purchase would be impractical, if not impossible. Instead, few of the more prominent variables suggested by economic theory should be included. Economic theory suggests *income*, in addition to price, affects consumption. The null and alternative hypotheses for the slope of income are as follows:

$$H_0: \beta_{income} = 0 \qquad H_1: \beta_{income} > 0$$

The results, after the inclusion of variable I_{Mean}, are shown in Table 4.3.

The result is much better. First, the model is significant. The F statistic has a p value (*Significance F*) of 0.024 indicating that there is 2.4% chance of committing a type I error if we reject the null hypothesis:

$$H_0: \text{Model is not good} \qquad H_1: \text{Model is good}$$

If we reject the above null hypothesis based on the outcome of the F statistics we would be wrong about 2% of the time. This is a low enough chance to take, so we reject the null hypothesis. The estimated equation is thus as follows:

$$\widehat{Q_{Beer}} = 22.1 - 4.7\,P_{Beer} + 0.00009\,I_{Mean}$$

The coefficients for P_{Beer} and I_{Mean} are both significant at 0.01 and 0.008 p value levels, respectively. If we reject the null hypothesis that the slope for the price of beer is zero, we would be wrong once in 100 times, while the probability of being wrong for rejecting zero slope for mean income is only 0.8%. Because these are low enough probabilities, the null hypotheses of zero slope should be rejected for both variables. Both estimated slopes have their theoretically expected signs. The slope of −4.7 for beer price indicates that a 10 cent increase in the price of beer would reduce beer consumption by 0.47 gallon per capita per year.

Table 4.3. *Output of Regression of* Q_{Beer} *on* P_{Beer} *and* I_{Mean}

SUMMARY OUTPUT

Regression statistics

Multiple R	0.659863279
R Square	0.435419547
Adjusted R Square	0.348561016
Standard Error	0.240325067
Observations	16

ANOVA

	df	SS	MS	F	Significance F
Regression	1	0.579060021	0.28953001	5.012973872	0.024334249
Residual	13	0.750829793	0.057756138		
Total	15	1.329889814			

	Coefficients	Standard error	t stat	p-value	Lower 95%	Upper 95%	Lower 95.0%	Upper 95.0%
Intercept	22.06207455	0.477759405	46.17821083	8.3861E-16	21.0299381	23.09421099	21.0299381	23.09421099
P_{Beer}	-4.704540094	1.598332187	-2.943405714	0.011415577	-8.157526848	-1.25155334	-8.157526848	-1.25155334
I_{Mean}	9.09356E-05	2.885E-05	3.152019164	0.007641872	2.86091E-05	0.000153262	2.86091E-05	0.000153262

A dollar increase in the mean income of Americans would result in a 0.009 cent more expenditure on beer. This might seem too small but it is not. Considering the fact that most people buy hundreds of items each year and many of them cost more than a beer, it is not too little to spend an extra 0.01 cents of an additional dollar of income on beer. However, these are normative issues related to one's taste. The fact is as stated above, regardless of whether it is too much or too little.

It might be tempting to compare the magnitude of the coefficients. The slope of the price of beer is 4.7/0.00009 = 52,200 times larger than the slope of the income. *This comparison is incorrect.* The issue is the magnitude of the unit of measurement. While the average of mean income for the period under study is $58,908, the average price of beer is $1 per 16 ounces. By changing the magnitude of the units of measurement, these values are changed without any impact on the estimation or prediction outcome. In fact, in most of the published research works, the units of measurement are changed before performing regression analysis to make sure that the coefficients have a similar number of decimal place. These changes make discussion and interpretation of results easier, but do not change the outcome of the analysis. Another factor that makes the above comparison meaningless is the fact that the corresponding *standard errors* are ignored. The only way the comparison could be meaningful is by calculating their respective Z score; in other words, normalizing the coefficients. This misinterpretation is more serious than the previous one.

The $R^2 = 0.4354$ means that about 44% of variation in beer quantity consumed is explained by the regression model that includes price of beer and average income. It is tempting to state that as R^2 for regression of beer quantity consumed on beer price (first model) was 0.0039, the difference 0.4318 or 43.18% of variation in the quantity of beer is explained by mean income. However, a simple regression of Q_{Beer} on I_{Mean} will show that the $R^2 = 0.0592$. In other words, the mean income alone explains about 5.92% of variation of Q_{Beer}. Had we started by regressing Q_{Beer} on I_{Mean} and then added P_{Beer}, we might have been tempted to say that 0.4354 − 0.0592 = 0.3762 or about 38% of Q_{Beer} must have been explained by P_{Beer}. The fact is that the P_{Beer} and I_{Mean} jointly explain about 44% of variation in Q_{Beer}, while individually they explain much less. It makes a difference only if one variable was in the model. Also, it makes a difference which variable is added first and which one is added second. The outcome of regression of Q_{Beer} on I_{Mean} is left for the reader as an optional additional practice. By regressing Q_{Beer} on I_{Mean} you will notice that the model is not even statistically significant. This issue arises in part from the fact that the two exogenous variables are not independent of each other as required. In fact, the correlation between P_{Beer} and I_{Mean} is 96%. You can verify this by typing the following in a cell:

$$=CORREL\ (range1, range2)$$

Replace range1 with the range for P_{Beer} and range2 with the range for I_{Mean}. When exogenous variables are correlated, especially as high as in this example, the test statistics are not reliable. For more detail refer to *Regression Pitfall* in Chapter 9.

As stated in Chapter 5 on *Goodness of Fit*, when a model has more than one independent variable, it is necessary to use Adjusted R^2 instead of R^2.

Estimating Y

In order to predict the value of an endogenous variable, we need to have specific values for independent variables. Insert the particular set of values for exogenous variables (X's) of interest in the estimated equation. For cross-section data, it is customary to use the average value of each X to find the average value of the response. This is the most reliable estimate because the input and output represent the center of the data with the least margin of error. In fact, it can be shown that the confidence interval for estimated exogenous variable $\hat{Y}$ is the narrowest at the center of the data.

However, this procedure is of little value for time series data because the average of the data is most likely in the distant past, while the future values are more useful. The problem with estimating future values of endogenous variable is that future values of exogenous variables are not known and because they are exogenous, they cannot be estimated either. There are two remedies for this. One is the use of a what-if analysis when we are dealing with policy issues. For example, it would be interesting to know what would happen to the quantity of beer if the mean income is the same as the last available data, that is, 2010 observation, or when the price of beer remains constant, but the government levies a 5-cent tax per 16-ounce bottle of beer. An alternative approach is to exclude the most recent available data, run a regression on the remaining data and then enter the unused exogenous variables in the estimated equation to estimate the last observed endogenous variable for comparison.

Example for Estimating Y

First, let us estimate $\hat{Y}$ using the last available data values of beer price and mean income. The price of beer for 2010 is \$1.23 and the average income for the same year is \$67,530. The estimated value of the quantity of beer is

$$\hat{Y} = 22.06 + (-4.7) \times (1.23) + (0.00009) \times (67,530) = 22.35$$

The difference between the observed value and the expected value of the Q_{Beer} is $22 - 22.35 = -0.35$ gallon per capita per year. This is the individual estimated error.

A 5-cent tax increases the price to \$1.28, and the estimated quantity of beer is

$$\hat{Y} = 22.06 + (-4.7) \times (1.28) + (0.00009) \times (67,530) = 22.12$$

A 5-cent tax increase, other things equal, would result in $22.35 - 22.12 = 0.23$ gallon reduction in quantity of beer consumed per capita per year.

Second, let us exclude the data for the year 2010 from regression. We leave the exercise to the reader. The estimated equation using data from 1995 to 2009 is

$$\widehat{Q_{Beer}} = 21.75 - 2.36\, P_{Beer} + 0.0000569\, I_{Mean}$$

Inserting the values of P_{Beer} = \$1.23 and I_{Mean} = \$67,530 for the year 2010, we obtain

$$\widehat{Q_{Beer}} = 22.68$$

The error or estimated residual is $22 - 22.68 = -0.68$ gallon per capita per year. The estimated value is fairly close to the actual observation and the error is "relatively small." However, this is a normative statement. The best way to determine the magnitude of the estimated residual is to calculate the standardized residuals.

Some Common Mistakes

The most common mistake using regression analysis is to let the data determine which variable belongs to the model. As you have seen so far, performing regression analysis is easy, yet without a theory, regression results can be misleading. Without economic theory, one might be tempted to regress independent variables one at a time and choose the one with the highest R^2 for inclusion in a multiple regression, then add the variable with the next highest R^2, until the resulting multiple regression outcome has desirable properties such as goodness of fit, high R^2, and significant coefficient.

We regress Q_{Beer} on each independent variable separately, and the results of R^2 are reported below.

X	R^2
P_{Beer}	0.0042
I_{Mean}	0.0597
Q_{Wine}	0.0357
Q_{Liquor}	0.0000
I_{Median}	0.0545

Note that none of the R^2 is high enough to be considered meaningful. We saw earlier that the P_{Beer} and I_{Mean} jointly provide a much better outcome, which is due to the high correlation between P_{Beer} and I_{Mean}. This high correlation is called *multicollinearity* and is addressed in more detail in Chapter 9 when discussing pitfalls of regression analysis.

Of course, one can take the above-mentioned incorrect idea to the next step by running all the possible two-variable regressions and compare their R^2s. There are 10 such possibilities. By using such approach, the next "logical" step would be to do all possible three-variable combinations, of which there are 10 possibilities. Then the set of all four-variable models could be run where there are five such possibilities. Finally, the set of all five variables can also be included, and its R^2 can be compared to the rest. The fact that these 25 different regression models have different numbers of exogenous variables means that at least the adjusted R^2 should be used instead of the R^2 to account for the differences in the degrees of freedom. Regardless, as there is no theory to govern the structure of our model, the practice is futile. Minitab software can perform all of these regressions with a single line of command. Another issue that adds to the uselessness of the above practice is that there is no theory, economics or otherwise, that could establish a link between the quantity consumed of one good and the quantity consumed of another good. Furthermore, it would make little sense to include both mean income and median income as explanatory variables in the same model.

Misspecification

The problem addressed in the last paragraph of the previous section is more serious than not using an economic theory. At most, only 1 of the above 25 possible models can be the correct model. The rest would have the wrong variables, too many variables, too few variables, or inappropriate variables. The exclusion of relevant variables, that is, exogenous variables that really affect the endogenous variable Y, results in slopes that are *biased*. This is known in the literature as *omitted variable bias*. It is not easy to calculate the extent of bias, because it requires knowledge of the variables that belong to the model and are excluded. If one knew of the variable and had access to the data, then it would not have been excluded. Furthermore, the magnitude and the direction of the bias depend on the correlation of the omitted variable with included variable(s), another unknown entity. Therefore, it is important to make sure that there is no omitted variable. One good way of doing so is to rely on economic theory. Another approach is to use econometric methods that are available for checking the omitted variables. There are some methods of overcoming the problem of omitted variable, but they are beyond the scope of the present work.

Another possible misspecification is the inclusion of variables that do not belong to the model. Theoretically, this is the least damaging of all problems. The expected value of the exogenous variables that do not belong to the model is zero. Furthermore, they do not have any impact on the slopes or the standard deviations of the estimates of other relevant variables. They do not have any effect on error either. Sometimes, these facts are used as the basis of throwing everything into the model and then deleting the variables that are not significant. The process is usually done one variable at a time. The variable with the highest p value is deleted, and a new regression is performed. This process, which is also automated in almost every software package, is known as *"backward elimination."* However, this method is also without foundation and should not be practiced. The "logic" behind eliminating the variable with the highest p value is the fact that if the null hypothesis of zero slope is rejected for that variable, it would result in the highest level of type I error. However, as variables are eliminated, it becomes evident that some of the variables that seemed to be significant before become insignificant after the elimination of undesirable variable(s). Of course, we know that the reason is the correlation among the included variables, as is discussed in Chapter 9 under section *Multicollinearity*. This raises the possibility that some of the variables that have already been excluded may become significant after some other variables are excluded. The fact is that some of the previously deleted variables will become significant if they were reintroduced later. A long time ago the process of backward elimination and reentering the excluded variable in later stages was automated into a procedure named *stepwise regression*, which is also available in most statistical software packages. This procedure also lacks theory, and there is no justification for the resulting model. The outcome is most likely due to fitting variables into specific data available. As soon as new sets of data become available, the same model obtained by the stepwise regression will produce poor results. Almost all such models are meaningless and ineffective in prediction.

Do not be fooled by "significant" slopes. Contrary to common wisdom, it is possible to use X values that do not have any relationship to Y and obtain "reasonable" results in the sense of goodness of fit, high R^2, and significant slopes. This is commonly known as *spurious regression*, which is discussed in more detail in Chapter 9 on *Pitfalls of Regression Analysis*.

Multiple Regression in Excel

Multiple regression involves the inclusion of more than one independent variable. Performing a multiple regression is similar to the above example with one exception. The columns of the independent variable must be adjacent or contiguous as it is commonly stated. So bundle all independent variables together in columns next to each other. The order, however, does not matter. When choosing the "Input X Range," use the same procedure as before, give the beginning cell address, and the ending cell address. Except now the range covers several columns instead of one.

For practice do the following:
Open a new spreadsheet. Repeat the above example from the beginning. However, instead of entering two (2) for the "Number of Variables" on the Random Number Generation dialog box, enter five (5). Insert a new row at the top and in the now empty cell A1 type "Dependent," and in cells B1, C1, D1, and E1 enter reasonable names for the independent variable names. For this example Ind1, Ind2, Ind3, and Ind4, might be reasonable.

Remember to adjust your entry for "Input X variable" to cover all four independent variables. The third part of the output will have five rows instead of two; one for the intercept, and one for each of the four independent variables, which now appear with their names.

No further rows or columns will ever appear in regression results. The only difference in the output of a simple regression with only one independent variable and that of a multiple regression is that the output of the latter will have additional rows, one for each additional independent variable. The entries for these rows will be identical to the ones for one independent variable, and their interpretations are the same as well, as is explained in the appropriate chapters.

An Example of Multiple Regression

Performing a multiple regression in Excel is exactly the same as performing a simple regression. Let us regress the quantity of beer consumed in gallons in the United States with the price of beer in dollars and the average income in dollars. The regression model is

$$Q_{Beer} = \beta_0 + \beta_1 P_{Beer} + \beta_2 I_{Mean} + \varepsilon,$$

where Q_{Beer} is the quantity of beer consumed, P_{Beer} is the price of beer per 16-ounce container, I_{Mean} is the average annual income, epsilon (ε) is the random error term, and the betas are the parameters of the model. Let us assume that column A contains the years 1995–2010,

Table 4.4. Beer Data

Year	Consumption (gallon per capita)	Prices/16 oz	Average annual income
1995	22.47	0.81	44,938
1996	22.51	0.84	47,123
1997	22.37	0.84	49,692
1998	22.52	0.86	51,855
1999	22.74	0.88	54,737
2000	22.72	0.92	57,135
2001	22.87	0.96	58,208
2002	22.98	0.99	57,852
2003	22.79	1.01	59,067
2004	22.93	1.07	60,466
2005	22.70	1.09	63,344
2006	23.00	1.11	66,570
2007	23.07	1.12	67,609
2008	22.98	1.16	68,424
2009	22.34	1.21	67,976
2010	22.00	1.23	67,530

Source: Beer Institute, Brewers Almanac 2011: PER CAPITA CONSUMPTION OF BEER BY STATE 1994–2010. US Census Bureau, Table H-6: Regions—All Races by Median and Mean Income: 1975 to 2010. Bureau of Labor Statistics, Consumer Price Index—Average Price Data: Malt beverages, all types, all sizes, any origin, per 16 oz.

while Q_{Beer} is in column B, P_{Beer} is in column C, and I_{Mean} is in column D. Row one contains labels for each column. The fact that consumption is reported in gallons and the price is for 16-ounce containers only affects the magnitude of the coefficient and not its significance. Make sure to pay attention to the units of price and quantity and report them correctly.

To perform a regression, execute the following functions:

$$Data \rightarrow Data\ Analysis \rightarrow Regression \rightarrow OK$$

In the Regression dialog box, enter the appropriate values as indicated below:

Input Y Range	B1:B17
Input X Range	C1:D17

Check the box on the left side of the "Label" to let the software know the first row is a text containing variable names and not data, then press OK. The results, shown in Table 4.4, are provided in a new worksheet.

Note that the only difference between the above instructions and the ones for the simple regression is that Input X Range contains both columns of exogenous variables. This means that in Excel all exogenous variables must be contiguous, which means they must be next to each other, so the range can form a rectangular shape. This limitation does not exist in most statistical software packages.

Descriptions of sources of each output and their sources are similar to their counterparts from the simple regression example.

Table 4.5. Quantity of Beer Regressed on Price of Beer and Average Income for the Years 1995–2010

SUMMARY OUTPUT

Regression statistics	
Multiple R	0.659863279
R Square	0.435419547
Adjusted R Square	0.348561016
Standard Error	0.240325067
Observations	16

ANOVA

	df	SS	MS	F	Significance F
Regression	2	0.579060021	0.28953001	5.012973872	0.024334249
Residual	13	0.750829793	0.057756138		
Total	15	1.329889814			

	Coefficients	Standard error	t stat	p-value	Lower 95%	Upper 95%	Lower 95.0%	Upper 95.0%
Intercept	22.06207455	0.477759405	46.17821083	8.38611E-16	21.0299381	23.09421099	21.0299381	23.09421099
P_{Beer}	−4.704540094	1.598332187	−2.943405714	0.011415577	−8.157526848	−1.25155334	−8.157526848	−1.25155334
I_{Mean}	9.09356E-05	2.885E-05	3.152019164	0.007641872	2.86091E-05	0.00015262	2.86091E-05	0.00015262

CHAPTER 5

Goodness of Fit

Testing the Goodness of the Model

The first and most important step in any empirical research is to determine whether the model used in the study is acceptable. When conducting an empirical study, there are steps that must be taken before starting the study and results to be checked after the regression is performed.

The foundation of any research in economics is economic theory. The importance of using economic theory and the proper procedure for conducting research have been addressed throughout this book. This chapter, which addresses one of the steps in empirical research, is about determining whether a particular regression provides acceptable results. In statistical analysis, the goodness of fit is determined by testing the following hypothesis:

$$H_0: \text{Model is not good} \qquad H_1: \text{Model is good}$$

Although all decisions based on statistical inference are normative, the statements in the null and alternative hypotheses seem more normative than typical statistical inference. Inferential statistics are normative in nature because the researcher makes a normative decision for rejection or failure to reject a null hypothesis based on the probability of committing a type I error. The selection of the level of tolerance for type I error depends on one's ability or willingness to take a certain level of risk; therefore, inferential statistics is normative. If the p value, which represents the probability of type I error, is "low enough" then the researcher rejects the null hypothesis in favor of the alternative hypothesis. Otherwise, one would fail to reject the null hypothesis. To make this test comparable to other inferential tests, we use the following criterion.

Rule 5.1

If the portion of the variation in the dependent variable that is explained by a regression (*MSR*) model exceeds the portion of variation that is not explained (*MSE*), then reject the null hypothesis.

Of course, the "numerical" difference has to be gauged against a norm or a standard, which is provided by an *F* distribution. Tabulated *F* values are provided by all software packages that perform a regression analysis. In Excel, this value appears under the heading of "*F*" in a table labeled "ANOVA."

A Brief Explanation of F Statistics

Theorem 5.1

Let U and V be independent variables having chi-square distributions with r_1 and r_2 degrees of freedom, respectively. Then the ratio

$$F = (U/r_1)/(V/r_2) \qquad (5.1)$$

has an *F* distribution with r_1 and r_2 degrees of freedom. A brief description of chi-square and *F* distribution is provided in Chapter 6.[1]

Theorem 5.2

If the random variable X has a normal distribution function with mean μ and variance σ^2, where variance is greater than zero, then the random variable $U = (X - \mu)/\sigma^2$ has a chi-square distribution with 1 degree of freedom, shown by $\chi 1$.[1]

Lemma 5.1

If the random variable U has a chi-square distribution function with 1 degree of freedom, then the sum of (r) many of it ΣU has a chi-square distribution with r degrees of freedom.[1]

These theorems and lemma explain the need for requiring the sample to be either drawn from a normal distribution or follow the requirements of the central limit theorem. Without an F statistics, it is impossible to test the goodness of fit of a regression model. Once the requirements and relations are met, the resulting F statistics can be used to test the goodness of fit of a regression model.

The table of analysis of variance, commonly displayed as ANOVA in most software, contains information that represents variables that are needed to obtain an F statistics. If the error terms in a regression model are random variables from a normal distribution with a given mean and a constant variance, then the *mean square regression (MSR)* will have a chi-square distribution with $k - 1$ degrees of freedom. In the same way, the *MSE* will have a chi-square distribution with $n - k$ degrees of freedom, where "k" refers to the number of parameters and "n" represents the number of cases, or sets of observations, which are represented as rows in data. Parameters of a regression model are its slopes plus the intercept. Therefore, a regression model with k-independent variable has $k + 1$ parameters; k parameters representing the k slopes associated with each of the k-independent variables plus the parameter representing the intercept. Sometimes *MSR* is called *explained variance*, while *MSE* is called *unexplained variance*. This terminology is somewhat different than the one used in this text where *variance* is the sum of squares of individual errors and individual errors are the *unexplained* deviations from the expected value.

F statistics have two degrees of freedom associated with them; the one listed first is for the numerator, which is the same as the degrees of freedom listed for regression in the ANOVA table. The one listed second is for the denominator, which is the same as the degrees of freedom listed for the residual line in the ANOVA table.

F statistics are used in the same way as any other statistics. The null hypothesis is rejected in favor of the alternative hypothesis if the calculated F statistics are *large enough*, which is the same as saying the probability of type I error or p value is *low enough*. P value is listed under "Significance F" in the Excel output in the ANOVA table. The null and the alternative hypotheses that are tested by an F statistics are as follows:

H_0: Model is not good H_1: Model is good

It is also important to understand what it means to say "model is not good." The simplest case is when the regression model has only one independent variable. In Chapter 2, we provided two different "lines" for estimating and predicting the dependent variable. The first line was simply the mean of the dependent variable, disregarding the existence of an independent variable. The second line was the use of regression model, which claims that at least part of the variation in the dependent variable is explained by the regression model. Recall that the total variation of the observations for the dependent variable is provided by the sum of squares total (SST), which is divided into two parts, that is, the part that is explained by regression model (Regression SS) and the part that is not explained by regression model (Residual SS). The part explained by regression model is estimated by the sum of squares of regression (SSR), while the part not explained by the regression model is estimated by the SSE. F statistics test the claim that the part of variation explained by the regression is substantially greater than the portion not explained by the regression; that is, SSR is relatively large compared with SSE. In order to put things in context, each part is averaged to account for the number of observations used to obtain them; that is, MSR and MSE are used. Needless to say, one could add variables to a regression model until the explained portion is sufficiently larger than the unexplained portion. In the extreme case, 100% of the variation in the dependent variable can be explained by the independent variables if there are $n - 1$ explanatory variables in the regression model; a perfect fit, albeit, is useless.

The above statement indicates that if the ratio of MSR to MSE is sufficiently larger than the numeral 1 to the point that the difference cannot be attributed to random error, then the model is doing a "good enough" job, and thus, the null hypothesis is rejected. Note the assumption that the error terms have a normal distribution is crucial. Without this assumption, the resulting ratio of MSR to MSE will not have an F statistics, and therefore, cannot be used to test the goodness of fit of the regression model.

The Case of One Independent Variable

Let us continue working with the simple linear regression of one independent variable. In the previous paragraph we referred to two models that can be used as regression model; one with and one without an independent variable. They are represented as equations (5.2) and (5.3), respectively.

$$Y = \beta_0 + \beta_1 X + \varepsilon \qquad (5.2)$$

$$Y = \mu_Y + \varepsilon \qquad (5.3)$$

The following theorem can be proven easily.

Theorem 5.3

The expected value of intercept of a regression line when slope is zero is equal to the mean of the dependent variable.

Therefore, testing the hypothesis that the model is not good, in the case of one independent variable, is the same as testing the hypothesis that the slope is zero. That is

$$H_0: \beta_1 = 0$$

This null hypothesis indicates that model (5.2), the regression model with one variable, is *not* correct and model (5.3) is correct. Model (5.3) indicates that no other factor explains Y and that its best estimate is its own average. This had been the unstated claim of introductory statistics where the expected value of a variable such as Y is its mean. The analogous graphical relation between the two hypotheses is easy to show. A line with a zero slope is a flat line parallel to the x axis. The graphical representation of the average of the dependent variable is also flat and horizontal as explained in Chapter 1. Because the estimates of both lines when the slope of the regression line is zero are exactly the same and equals the sample mean of the dependent variable, then the two lines coincide as in Figure 5.1

The hypothesis about slope can be tested using a t statistics with $n - 1$ degrees of freedom as well. As it is seen, in the case of a single independent variable, whether one uses an F statistics or a t statistics is irrelevant because they are equivalent and produce the same result. This is not

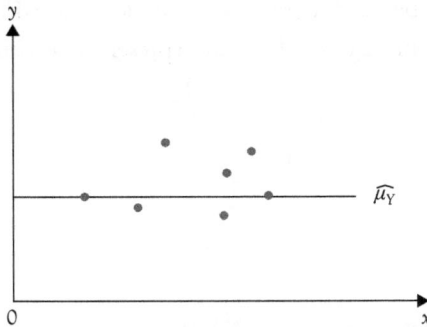

Figure 5.1. Regression equation when the slope is zero and the average of the dependent variable.

a mere coincidence. There is a relationship between F and t when the degree of the freedom for the numerator of the F distribution is only 1.

Theorem 5.4

The F statistics with 1 and $n - k$ degrees of freedom is equal to the *square* of a t statistics with $n - k$ degrees of freedom.

$$F = t^2,$$

where "n" is the number of observations and "k" is the number of parameters.

Remember, if p value is low enough, we reject the null hypotheses and report the level of significance, that is, the level of type I error.

The Case of Two or More Independent Variables

The above statement regarding the use of F statistics for testing goodness of fit of the model can be applied to multiple regression as well. Null and alternative hypotheses for testing goodness of fit for multiple regression model as shown in equation (5.4)

$$Y = \beta_0 + \beta_1 X_1 + \beta_2 X_2 + \cdots + \beta_K X_K + \varepsilon \qquad (5.4)$$

are the same as for the case of simple regression:

H_0: Model is not good H_1: Model is good

However, the null hypothesis that "model is not good" now is more general and applies to all slopes. An alternative way of representing the same null hypothesis is the complex hypothesis involving all the slopes of the model:

H_0: $\mu_1 = \mu_2 = \cdots = \mu_k = 0$ H_1: At least one mean is different from zero

Because this is a complex hypothesis, it *cannot* be tested with a t statistics, rather it requires an F statistics to test. It would be incorrect to conduct K different t tests to test the above hypothesis. The only reason for presenting these equivalent null and alternative hypotheses here is to familiarize the reader with the notation, which might be encountered in the literature. Remember, if p value is low enough, reject the null hypotheses and report the level of significance, that is, the level of type I error. Regardless of how the null hypothesis is stated, the result is the same, which is stated in Theorem 5.3, which is an extension of Theorem 5.1.

Theorem 5.5

The expected value of intercept of a regression line when *all* the slopes are zero is equal to the mean of the dependent variable.

Under the null hypothesis of "model is not good" the multiple regression model (5.4) collapses to model (5.3), which indicates that the best estimate of the values of dependent variable when regression model (5.4) is not valid is their mean, which is μ_Y.

What to Do After Testing the Goodness of the Model?

The course of action after the completion of the F test in testing the goodness of the model depends on researcher's decision. If the researcher finds the probability of type I error prohibitively too high to reject the null hypothesis, he or she must "fail to reject" the null hypothesis. This would be the end of the study. There is no reason to proceed any further because the proposed model failed to provide sufficient evidence to refute the existing expectation.

If the test of hypothesis is rejected, there are two options that can be taken based on one's preference, or on the objective of the research. One is to check the coefficient of determination, commonly known as R^2 and pronounced R-squared. The second option is to examine each of the coefficients or the slopes for significance and provide inference and analysis. We cover the latter part in Chapter 4 under section *Interpretation of Coefficients*. For now, we tend to the first option. The important thing to remember is that although there is a logical starting point, namely conducting a test of goodness of fit with an F test, there is no other single logical path because there is no sequential path for regression analysis. Everything must be examined and utilized in its entirety in order to obtain a realistic view of what the data reveal about the model and whether the evidence supports the claims of the research.

Coefficient of Determination or R^2

The method of least squares provides the best unbiased estimated line for a given data. The best is used in the context of efficiency; that is, the estimator has the smallest variance in the class of unbiased, linear estimators. This, however, does not provide any information on how good the regression line fits the data. It does not mean that the model is "good enough" for any particular purpose, thus requiring further examination and inference.

The regression line explains variations among the values of the dependent variable beyond what the mean of the dependent variable can explain, using independent variable(s). A measure of this ability is called *coefficient of determination*. It consists of the ratio of *SSR*, the portion of variation explained by regression, to *SST*, total variation in the dependent variable.

$$R^2 = SSR/SST$$ (5.5)

Customarily, it is expressed in terms of a percentage. It ranges between zero (0) and one (1) or (0%) and (100%), depending on the expression choice. Higher values of R^2 indicate higher explanatory power of the model. However, there are situations when a high R^2 does not necessarily mean a better fit.

Adjusted R^2

Theoretically, the magnitude of R^2 should improve only with improvement in the ability of the model to explain larger share of total variation in dependent variable. However, if additional independent variables are included in the model, even if they are only random numbers with no explanatory powers, the magnitude of R^2 will increase. This is due to two factors. First, the value of R^2 can never decrease with the inclusion of additional exogenous variables; it can either remain constant or increase. The value of R^2 would remain the same after a new variable is included if, and only if, the new variable has no explanatory power and that it is also independent from all other exogenous variables already in the model. Because the probability of two exogenous variables having a correlation coefficient of zero is very low in social sciences, inclusion of additional exogenous variables customarily causes an increase in R^2. Earlier it was stated that it is possible to provide a perfect fit for any dependent variable using $n - 1$ independent variables in the model. In other words, with $n - 1$ exogenous variable R^2 would approach 100%, even if some of the independent variables have no explanatory power. To solve this problem, a formula, called *Adjusted R^2*, is used to correct for the loss of degrees of freedom caused by including more independent variables. In Excel, Adjusted R^2 appears on the third row of the "summary output" in upper left-hand corner. When dealing with two or more independent variables, it is better to use, especially for comparisons, the adjusted R^2 when dealing with multiple regression.

The Difference Between R^2 and SSR

Sum of squares due to regression, *SSR*, represents the amount of variation of data points in dependent variable explained by regression model.

R^2 is the proportion or the percentage of variation of data points in the dependent variable explained by regression model. SSR can take any positive value. It can be large or small. The magnitude of SSR changes with the unit of measurement. R^2 is a percentage or proportion between zero and one. R^2 is immune to changes in units of measurement.

Relation Between R^2 and ρ

In simple regression, where there is only one independent variable, coefficient of determination (R^2) is equal to the *square* of the correlation coefficient (ρ) between dependent and independent variables. From a sample, the following could be verified.

$$R^2 = (r)^2, \tag{5.6}$$

where r is the *sample* correlation coefficient. The sign of r is the same as the sign of β_1. For a multiple regression model, the coefficient of determination is not equal to the square of correlation coefficient due to the interaction between independent variables and due to the presence of more than one independent variable. There is a relation between R^2 and all the partial correlation coefficients in the model, usually presented in more advanced courses.

Relation Between R^2 and F

The coefficient of determination (R^2) is related to F distribution through the following formula:

$$F = \frac{R^2}{1 - R^2} \times \frac{n - P}{P - 1}. \tag{5.7}$$

The relation can also be expressed in terms of R^2, instead.

$$R^2 = \frac{(P-1)F}{(P-1)F + (n - P)}, \tag{5.8}$$

where R^2 is the (multiple) coefficient of determination, n is the sample size, and P is the number of parameters. Note that there is one more parameter in the model than there are independent variables $P = K + 1$.

As R^2 increases F increases, as P increases F statistics decreases. Inclusion of additional independent variables will increase P but may or may not increase R^2. If the added variable does not belong to the model, then F statistics might become insignificant.

CHAPTER 6

Regression Coefficients

A regression model should reflect reality. In economics, reality is explained by economic theory. Therefore, a regression model about economics should be based on economic theory, and the results must also be explained in terms of economics.

Coefficients of Simple Regression

Estimating a Consumption Function

The simplest form of consumption theory in economics states that "consumption is a function of income." This macroeconomic relationship is written as

$$\text{Consumption} = f(\text{income}) \tag{6.1}$$

This equation is read as "consumption is a function of income." The simplest functional form and the one most commonly used is the linear function.

$$\text{Consumption} = \text{subsistence consumption} + (\text{marginal propensity to consume}) \times (\text{income})$$

This economic theory can be written in the symbolic form as

$$C = \beta_0 + \beta_1 Y, \tag{6.2}$$

where C represents "consumption," (Y) represents "income," β_0 is the subsistence level of consumption when income is zero, and β_1 is the marginal propensity to consume. Algebraically, β_0 and β_1 represent the intercept and the slope of a straight line that presents consumption as a linear function of income. Equation (6.3) is considered an economic model because it is a simplification of reality. However, it is not a regression model as will be explained shortly. As stated before, equations similar to (6.3) are not statistical models due to the fact that they lack an error term. A regression model also includes a random error term, which is customarily represented by the letter epsilon (ε).

$$C = \beta_0 + \beta_1 Y + \varepsilon \qquad\qquad (6.3)$$

This regression model can be tested for goodness of fit, which would actually test the economic theory. The *Goodness of Fit*, as stated in Chapter 5, is tested with F statistics. Economic theory also states that the *MPC* is between zero and 1; in other words $0 < \beta_1 < 1$. The trivial cases of equality with zero or one are not of interest. This aspect of the economic theory can be tested. In order to do so, two different tests must be performed. One which tests that *MPC* is less than one, and another to test that it is greater than zero.

$$H_0 : \beta_1 = 0 \qquad H_1 : \beta_1 > 0$$

The above hypothesis is tested using a t statistics. Note that the assumption that the error terms have a normal distribution is crucial. Without this assumption, we cannot use the t statistics to test that the slope is positive.

Testing the second part of the inequality requires a moment of reflection. The claim of the second part of the inequality is that the *MPC* < 1. The value that nullifies this claim is one, not zero, so the null hypothesis must be

$$H_0 : \beta_1 = 1 \qquad H_1 : \beta_1 < 1$$

The procedure for testing this hypothesis is similar.

An Example

Annual data from 1990 to 2010 on personal consumption expenditures and population are available from the Bureau of Economic Analysis. The data in the following site was assessed on November 28, 2011.

http://www.bea.gov/iTable/iTable.cfm?ReqID=9&step=1

Copy and paste the above link on your browser. Once the National Income and Product Account Tables are populated, click on *Section 2— Personal Income and Outlays* from the list of tables. Select *Table 2.3.5. Personal Consumption Expenditures by Major Type of Product (A) (Q)*. To change the data range to start from 1990, click on the "Options" icon that is located on the top right corner above the table. The "Options" icon will open a window. Change the Series to "Annual" and change the First Year to "1990" and click on Update. Line number "1" on the table represents the annual personal consumption expenditure. Or you can select the "Download" icon, which is in the same box as the "Options" icon. Choose the version of Excel, which is compatible with your version. Highlight the row of labels, containing the years and the first row of data, which pertains to annual personal consumption expenditure. Choose copy and open a new worksheet. In cell A1, click on the right button on your mouse; choose "paste special" and finally check the box "transpose" to get the data to display as a column instead of a row.

The per capita personal income from 1990 to 2010 is obtained from the following source on the same date.

http://www.bea.gov/iTable/iTable.cfm?ReqID=70&step=1

Copy and paste the above link on your browser. Once the GDP and National Income tables are populated, click on the *Annual State Personal Income and Employment*, and then click on the Personal income and population (SA1-3). Select the *SA1-3—Personal Income Summary* and click "Next Step." Select "United States" for the Area and "All Years" for Year, then from the drop-down list for Statistics select Per Capita Personal Income (dollars) and click "Next Step." You could select the "Download" icon, which is next to the "Options" icon to generate an Excel file. Highlight the row of labels and the first row of data, which pertains to annual personal consumption expenditure. Choose copy; open a new worksheet; choose "paste special" and finally check the box "transpose" to get the data to display as a column instead of a row. Now the dates and the data match. Save your work. The first two rows of data are depicted in Figure 6.1.

	A	B	C
1	**Year**	**Income**	**Consumption**
2	1990	17004	15331
3	1991	17532	15699

Figure 6.1. Income and consumption data.

Enter income in cells B1 and consumption in C1, then save your work. Make a note of the full name of data and their sources for future reference. The best practice is to keep the original download file as it is and save the modified data in a separate spreadsheet or just another worksheet. Label each worksheet accordingly for reference. Make sure income is in column "B" and consumption is in column "C." This assures that following the steps results in the same output as provided below. To perform a regression, execute the following commands:

Data → Data Analysis → Regression → OK

Enter the appropriate values in the drop-down boxes as indicated below:

Input *X* Range   B1:B22
Input *Y* Range   C1:C22

Click on the button on the left side of the "Label" to let the software know that the first row is a text containing variable names and not data. Click OK. The following result is provided in a new worksheet.

Analysis Regression Results for Estimating the Consumption Model's Coefficients

The beginning point for regression analysis, as explained in Chapter 5, is the test for the following hypothesis:

H_0: Model is not good H_1: Model is good

Table 6.1. The Results of Regression of Per Capita Personal Consumption Expenditures on Per Capita Income 1990–2010

SUMMARY OUTPUT

Regression statistics	
Multiple R	0.998548976
R Square	0.997100057
Adjusted R Square	0.996947428
Standard Error	340.6088128
Observations	21

ANOVA

	df	SS	MS	F	Significance F
Regression	1	757902463.9	757902463.9	6532.852199	1.41294E-2S
Residual	19	220266.433	116014.0228		
Total	20	760106730.3			

	Coefficients	Standard error	t stat	p-value	Lower 95%	Upper 95%	Lower 95.0%	Upper 95.0%
Intercept	−663.7179653	315.7179223	−2.102249883	0.0490951	−1324.52317	−2.912761022	−1324.52317	−2.912761022
Income	0.947876332	0.01172736	80.82606139	1.41294E-25	0.923330685	0.972421979	0.923330685	0.972421979

The null hypothesis that the "model is not good" can be written as a set of complex hypotheses in terms of all the slopes of the model:

$$H_0: \mu_1 = \mu_2 = \cdots \mu_k = 0 \quad H_1: \text{At least one mean is different from zero}$$

To test this hypothesis, examine the "significance F," which is the Excel jargon for the probability of committing a type I error, if the above null hypothesis is rejected. The listed level of significance is 1.41294E–25, which means that the decimal place must be moved 25 places to the left of where it is shown. This is extremely low, and we reject the null hypothesis that the model is not good in favor of the alternative hypothesis, which states the model is good. This means that at least one of the independent variables has a coefficient that is not zero. However, the model only has one independent variable, which by default must have a slope that is not zero. Nevertheless, it helps to go through the steps of interpretation of the results. Following the same procedures will ensure proper analysis and reinforce the practice.

It is a good habit to write the estimated equation. It is usually what is reported in a published work. Customarily, the calculated t statistics are reported in parentheses under the equation. A better practice is to report the p values instead.

$$\widehat{\text{Consumption}} = -663.71 + 0.95 \text{ income}$$
$$(0.049) \qquad (1.41\text{E}{-}25) \qquad\qquad (6.5)$$

The values in parentheses under the estimated equations are the p values for each estimated parameter. Some people prefer standard errors, others prefer t statistics, but this author prefers p values, which can be used directly and without the need for further computation or use of reference tables to make an inference. Some people would immediately proceed to explaining the meaning of the coefficients. This is a poor practice. The magnitudes of coefficients are really meaningless if one cannot reject the null hypothesis that they are equal to zero. The hypothesis about the slope is

$$H_0: MPC = 0 \quad H_1: MPC > 0$$

Or equivalently in symbolic form:

$$H_0: \beta_1 = 0 \qquad H_1: \beta_1 > 0$$

The p value associated with MPC is small, and we should feel confident in rejecting the null hypothesis. The probability of committing a type I error when rejecting this null hypothesis is a little over 1.4 in 10^{25}, which is equal to a 1.4 with 25 zeros to the right of it; in comparison, the population of the world is 7×10^9, which is 7 with only 9 zeros in front of it. Therefore, we should feel confident that rejecting the null hypothesis is the right thing to do. The conclusion is that MPC is not zero; it is positive. An observant student will notice that the probability of type I error for the t statistics for the slope of the independent variable, that is, income is the same as the p value for the F statistics. This is always true for the case of a single independent variable, as elaborated in Chapter 5.

After rejecting the null hypothesis that the slope is zero, it is possible to examine its meaning. The slope of 0.95 means that Americans will spend 95 cents of every additional dollar they earn and save the rest, which is 5 cents. A better way of looking at this aspect of the results is through a confidence interval. We chose not to change the default level of confidence, which happened to be 95% for Excel as default. The 95% confidence interval for the real MPC is between 0.92 and 0.97 in the above example; the range 92–97 cents includes the real MPC with 95% confidence. It is a coincidence that MPC is 95%, which is the chosen level of confidence.

Test of Hypothesis About Intercept

The procedure for testing a null hypothesis about intercept is exactly the same. If the probability of committing type I error is small, reject the null hypothesis; if not, fail to reject it. But what are the null and alternative hypotheses for intercept?

The simplistic consumption equations of (6.1) and (6.2), and therefore, their corresponding statistical model expressed in equation (6.4) indicate that if one's income is zero, he or she would consume an amount equal to β_0, which represents the subsistence level, at least in the macroeconomic theory of consumption. This implies that the subsistence level of income should be low and positive. In fact, the idea of a negative consumption is absurd. So a reasonable claim about the subsistence level is that it should be positive, or greater than zero. The number that nullifies this claim is zero. Therefore,

$$H_0: \beta_0 = 0 \qquad H_1: \beta_0 > 0$$

Once again, note that the process began with the alternative hypothesis. After establishing an appropriate alternative hypothesis based on some economic theory, the value that nullified it is stated as the null hypothesis. This test really does not require any computation. The claim is that the intercept is *positive*, but regression estimate in equation (6.5) is *negative*. Therefore, there is no evidence to support a positive intercept based on the data from the years 1990–2010. A negative number can never provide any support for a claim that the true level of subsistence is positive no matter how far it is from the center. Thus, we fail to reject the null hypothesis. If we reject the null hypothesis, the probability of type I error is a little over 50%. In fact, it is exactly:

$$0.5 + 0.04909 = 0.54909$$

Some Common Mistakes

The novice often refers to the output and observes a negative number in front of the estimated intercept and then chooses the following (*incorrect*) null and alternative hypotheses:

$$H_0: \beta_0 = 0 \qquad H_1: \beta_0 < 0$$

He/she proceeds by observing the reported p value of 0.04909, which is a small probability, and rejects the null hypothesis in favor of the alternative hypothesis. *This is incorrect* because there is no economic theory that explains a negative subsistence level.

The original claim is the correct claim, and the theory anticipates a non-negative consumption in spite of lack of income, which also makes common sense. However, there is not sufficient evidence to reject the null hypothesis that when one's income is zero his or her consumption is zero as well. In other words, according to the estimated equation, when income is zero one ceases to consume and dies. This is a plausible, although it is an oversimplification of reality, at least in the short run. In the short run, one would not die as a result of lack of income. Even with zero consumption, it would take a while for death to occur. Where the model clashes with reality is in the fact that the model ignores the possibility of having savings, wealth, family, relatives, and not to mention theft. The appropriate thing to do would be to improve the model by including the above-mentioned control variables in the model, rather than jumping to an illogical conclusion.

Another common mistake is to look at the magnitude of -663.717 (the coefficient of the intercept from Table 6.1) and conclude that it is large. Because measurements are in dollars, this number reflects a negative $663.71. Although not a fortune by any stretch of imagination, it is nevertheless a far cry from a zero dollar consumption of the null hypothesis. The fact is that in order to have an average = 0, one must observe some negative values and some positive values in such a way that the average is zero. In this example, it happens that the observed estimate is negative. The magnitudes of positive and negative values depend on the variance of the data. The more problematic issue is the fact that the estimate has a relatively small standard error as compared to the coefficient, namely 315. This results in a relatively large t statistic of -2.10. Although, as explained above, this evidence cannot be used to conclude a positive subsistence consumption level, nevertheless, it would make one uncomfortable stating that the intercept is zero. This is further evidence that when we *fail to reject the null hypothesis*, we need to be cautious and not conclude that the null hypothesis is accepted.

For this and other reasons to be explained later, there is little stock placed in interpretation of the intercept of a regression model in many instances. Instead, the main focus is on the slope, and whether it is zero. A zero slope indicates the failure of an independent variable to explain variation in dependent variable; indicating that inclusion of that particular variable does not improve the explanatory power of the model. Therefore, the best variable to explain variation in the dependent variable is its mean.

One should not form a claim, that is, an alternative hypothesis, based on empirical evidence from a sample. The claim must be made before the data are collected. Also, alternative and null hypotheses must be stated first. Otherwise, an error will be made by rejecting the null hypothesis using the incorrect alternative hypothesis, which is known in the literature as the *type III error*.

Definition 6.1

A *type III error* is rejecting a null hypothesis in favor of an alternative hypothesis with the wrong sign.

Coefficient of Determination: R^2

Coefficient of determination shows what percentage of variation in dependent variable is explained by a regression model. In this case, 99.88% of variations in consumptions are due to differences in income. This is an impressive performance for a humble model. It is important to know that when two variables increase overtime, they will do a good job explaining variations in each other. This occurs in time series data frequently. Therefore, other things being equal, R^2 would be higher for time series data, such as income and consumption from 1990 to 2010 in this example. The problem will be addressed in greater detail in Chapter 9, Spurious Regressions.

A low value of R^2 indicates, among other things, that a small percentage of changes in the dependent variable are explained by a regression model. However, a high value of R^2 can either mean a high explanatory capability or it could be due to other factors, one of which is addressed in the next paragraph. The terms "low" or "high" are indeterminate and vague. Because $0 \leq R^2 \leq 1$, the extreme values for low and high end are known. What is not known is how close the value of R^2 must be to zero before it is considered low, and how close it should get to one to be considered high. Statements about the explanatory power of R^2 are one of the rare occasions where no probability is provided for the accuracy of the statement. Therefore, they lack the precision of statements associated with F statistics, for which probabilities of type I and type II can be calculated and a power of a test determined.

Another problem with the use of R^2 is that it gives a point estimate and does not have a familiar or commonly used confidence interval. However, there is a relation between R^2 and F value as expressed in Chapter 5 under the section *Relation between R^2 and F.*

CHAPTER 7

Causality

Correlation Is Not Causality

Chicken or Egg?

Siblings usually have similar heights. Tall people are expected to have tall siblings, and short people are expected to have short siblings. At least this is the conventional wisdom. Which sibling's height explains the height of the other siblings? Is it the first child's height that sets the height expectations for the rest of the siblings or the last child's height? The question is whose height *causes* the height of the other siblings? Although most people would agree that sibling heights are highly correlated, few people would expect one child's height to explain a sibling's height. The cause, or the explanatory variable that determines sibling heights, is the heights of parents. However, a regression of height of one sibling on another sibling would result in a highly significant coefficient and other measures of suitability of the model. In this example, it is obvious that the source of high correlation between two siblings is not causal. The cause is the height of the parents.

How about correlation between income and consumption? Does income determine consumption, consumption determines income, or is there something else that determines both? The economic theory states that the causal direction must be from income toward consumption. As income changes, level of consumption changes because a fraction of each additional unit of income, known as the "marginal propensity to consume out of income," is spent on consumption. The direction of causality is logical. When A causes B, then A can occur without B occurring, but B cannot occur unless A occurs. This type of causality exists between parents and children. Similar, but not exactly the same, relationships exist between income and consumption. Typically, earning income entitles one to consumption, while the opposite is not true. Nevertheless, one can imagine consumption without income, such as through borrowing, charity, theft. The important thing is that consumption does not cause income.

Regression analysis can and does provide evidence for causality, but the reason for causality is rooted in economic theory, not in the results of regression analysis. Recall how in Chapter 1 a random variable was regressed on another random variable. Obviously, the random variable that was used as the independent variable did not cause the random variable used as dependent variable. Regression analysis provides evidence to disprove a hypothesis or fails to do so. Whether the hypothesis is false or groundless does not change the computations of regression analysis and, therefore, the outcome. Results of a regression analysis should never be used to form hypotheses about a model or its coefficients. In fact, claims about regression coefficients must be made, using economic theory, before any data are gathered.

The Role of Theory

Demonstrating a causal relationship requires a valid theory. The theory states the requirement and the outcome. For example, the theory of demand states when there is an exogenous increase in income, demand curve will shift to the right, and the price and quantity demanded will increase. Any time the theory's requirements are met, the theory's prediction would apply. This is a testable hypothesis using regression analysis. Note that the claim is that an increase in income causes the above-mentioned outcomes. Ignoring this causal factor and obtaining data on quantity demanded and price, and regressing the first on the second, would result in incorrect conclusion that a price increase causes an increase in the quantity demanded. Careless use of data and regression analysis without theory could prove disastrous. Demand theory is correct universally everywhere and for every good, except for an inferior good. An *inferior good* is a good with *negative income elasticity*; an increase in income would result in a decrease in quantity demanded for an inferior good. Therefore, the customary assertion that an increase in income causes an increase in quantity demanded does not apply to inferior goods. Most of studies using cross-sectional data can be causal in nature. In such studies, a theory is stated. The theory is believed to apply to all populations under study, such as firms, individuals, regions, or countries, and there is a causal effect between independent variable(s) and the dependent variable. If there are exceptions, such as the case of inferior goods in the above example they are noted. Such studies are suitable for verification of theories.

Direction of Causality

Causality is more evident in some studies than in others. For example, in a study of relationship between height of a father and height of his son, the direction of causality is clearly from father to son and never the other way around. The direction of cause-and-effect relationships between education and income is not as clear. It seems that people with higher levels of education should have higher levels of income, indicating a causal relationship where education determines income. However, it is easy to argue that people with higher (family) income attain higher levels of education, which places the causal relationship in the other direction, where income determines educational attainment. In fact, children of poor families have lower educational attainment, which in turn perpetuates poverty. A child's educational attainment cannot be the cause of low family income, while low family income can, potentially, cause low educational attainment of the children of the family. In this case, using family income instead of individual income clearly establishes the direction of causality. We tend to believe or hope that the direction is from education to income and not the other way around, but the reality and logic cannot be ignored. We are not, however, claiming education does not have a positive effect on income. Sometimes, through using lagged variables, one can establish the direction of causality. However, it might not be possible to use this methodology to test the possibility of education causing income, because among other things, income levels are *serially correlated*, which means this year's income is highly correlated with last year's income.

Sometimes the direction of causality is misunderstood because of poor comprehension of the theory. In economics, the theory of demand states that a change in price results in a change in quantity demanded in the opposite direction, except in the case of a Giffen good. A novice might argue that if the quantity declines, the price must increase. The novice is correct in that a decline in quantity would result in an increase in price, but the cause is not a decline in quantity. Rather the cause is what made the quantity to decline. The statement, as presented, is a statement about the quantity *supplied* not *demanded*. Furthermore, the supply would not decline by itself; something must cause the reduction in supply, such as a natural disaster, or change in production capacity. It is not wise to ignore the origin of a cause and only look at part of a cause-and-effect chain.

What if the subject of the study does not have an existing theory? The necessary data to conduct research based on a theory cannot be obtained. Similar problems exit when there are conflicting or contradictory theories, the assumptions or the requirements of the theory cannot be met, or the theory does not provide a specific or measurable outcome. Examples abound, but only few suffice to make the point clear. Stock market prices follow a random walk model, which is stochastic in nature and cannot be estimated with conventional models such as regression. Factors that affect weekly sales are nonlinear and are affected by the calendar date as well as seasonal effects, and therefore, cannot be estimated with linear models such as regression. When an inappropriate linear regression model is used, the outcome would be meaningless.

Association Without Causality

In many cases problems that render regression-based causal models ineffective can be remedied by other methods. For example, a good estimator or even a predictor of today's temperature is yesterday's temperature and the temperature of the same date a year before. An economic example is the price of a home, which depends on the price of the neighboring homes, location, and other hedonic factors rather than the economic theory; this does not mean that the economic theory of supply and demand does not have a role in determining the price of a home. The point is that it is somewhat meaningless to say that yesterday's temperature "caused" today's temperature. Although regressing the temperature of a day on the temperature of the day before results in a "good fit," it does not prove causality. Both the temperature of yesterday and today depend on the time of year, which changes the distance from the sun and the angle of the earth toward the sun. Here, the existence of the sun, the distance from it, and the earth angle toward the sun are the causes of a day's temperature and all other factors are control variables.

Many time series data present characteristics that do not render themselves to causal regression analysis. There are specific regression-type analyses that apply to such data but are beyond the scope of this book. In general, noncausal models, especially those that depend or use time series data, are reliable in the short run but not in the long run. Causal-based models are reliable in the long run but might be affected by short-run shocks and do poorer in the short run. When regression is applied to cases where there are no theoretically stated causal relationships, the results only indicate an association and not causation. Therefore, causality must be established before the regression is performed and not inferred from the outcome of a regression analysis. A famous historical example is the statement that the sun spot activities cause business cycles. A high degree of association, represented by a large R^2 value, does not indicate causation.

Ceteris Paribus

Students of economics are familiar with the concept of *ceteris paribus*, which means *other things being equal*. Economists have noticed long ago that numerous factors affect a particular phenomenon. For example, in demand theory, the relationship of interest is the link between price and quantity. In economics, the relationship is stated in terms of quantity being a function of price, which means a change in price would cause a change in quantity demanded. A change in quantity would also change price, however, the change in quantity is a supply issue, as stated earlier. This does not mean that nothing else affects quantity that one demands for a given price. Changes in taste, lifestyle, needs, family status, etc. all affect quantity demanded even if the price does not change. Imagine what would happen to quantity demanded of many goods for a student who graduates, finds a job, gets married, has children, etc. These are assumed to cause a shift in *demand schedule* while changes in price cause a change in *quantity demanded*. In order to create a demand theory, all of the other factors must be assumed to remain constant. In natural science, the factors that are not of interest can be kept constant while the same is not possible in social sciences. It would be unreasonable to tell a control group not to have children or finish their education so you can study their demand function. In regression analysis, these factors are included in the model and are called *control variables*. Although these variables are not of interest, nevertheless, their impact on the dependent variable must be accounted for and reported. Note that in multiple regression the contribution of each coefficient to the dependent variable is based on the presence of all the other variables in the model. In other words, each variable is treated as if it was the last variable to be added to the model. Therefore, β_i is the impact of one unit change in X_i given that all the other variables are already in the model and their effect has been accounted.

The study of causal relationships between two factors while keeping other things equal does not preclude the possibility of studying a causal effect of one of the control variables in its own right. For instance, in the study of demand, the causal relationship of interest is between price and quantity demanded. In the process it is necessary to keep income constant. However, the study of causal relationship between income and quantity demanded has been pursued in the literature with as much interest as the price–quantity relationship and is still covered in all microeconomics books. *Income–consumption* curve is also known as *Engel curve*. The slope of the Engel curve is used to group goods as inferior, normal, necessity, and luxury goods. In a study of an Engle curve the price of the good, among other things, is assumed to remain constant.

CHAPTER 8

Qualitative Variables in Regression

Qualitative Data

There are numerous types of data, each with its own characteristics and nature. Qualitative data are information about characteristics of a population of interest that cannot be measured numerically. For example, gender of workers is important in wage-earning differentials or comparisons of educational attainment of unemployed people. However, gender cannot be represented numerically; at least not in the same way as quantitative variables. Other qualitative characteristics are race, eye color, city or state of residence, type of car one drives, kinds of food one purchases, and numerous other characteristics that cannot be presented meaningfully with numbers. Qualitative variables have other names such as qualitative data, binary data, and dummy variables, which are all commonly in use, not only in economics, but also in other business fields and in social sciences. Types of variables are discussed under the *measurement scales*.[1]

It is possible, and in fact common, to represent qualitative variables as binary numbers, customarily zero and one. One outcome receives one value and the other outcome receives the other. For example, male can be represented as one (1) and female as zero (0), but there is no compelling reason why one or the other gender should be labeled as one (1). There are some advantages in using these values instead of say 1 and 2, or 14 and 37; although if the objective is to classify or group these variables, the result would be the same. While mathematical operations on quantitative variables are meaningful, they do not make any sense when applied to qualitative variables. This is one reason for the use of 0 and 1 instead of any other digits. These two digits do not have any proportional properties, while the "number 2" for instance is twice as big as the "number 1." Adding or subtracting zero to one does not result in different numbers and would not indicate any proportional value either. There are other advantages to doing so as well. Dummy variables have limitations too, which are addressed in this section.

In regression analysis, qualitative variables are represented in the form of combinations of binary values 0 and 1. The variable of interest, say gender, race, color, preference, etc. can be the purpose of the study, that is, the *dependent variable*, or the explanatory variable, that is, the *independent variable*. This chapter deals with implementing and interpreting the use of qualitative variables as an *independent variable* only, which is customarily called the *dummy variable*.

It is important to realize that there is a substantial difference in the required methods of incorporating qualitative information into a model and performing statistical analysis when the variables are the independent variable as compared to when they are the dependent variable. When the dependent variable is qualitative, the method of least squares is inappropriate and cannot be used because the *assumptions* listed in the appendix cannot be met. Several statistical methods are available to deal with cases where the dependent variable is qualitative in nature. Some examples of such methods are *logit, probit, discriminant analysis*, and *survival analysis*. These subjects and methods are beyond the scope of this chapter. This chapter deals with qualitative variables as independent variables. The names listed above are used interchangeably in this book.

Qualitative Independent Variables

Qualitative variables are represented by a series of *zeros* (0) and *ones* (1). The value of zero is assigned to one of the two categories of the qualitative variable, and the value of one is assigned to the other variable. If there were three males and two females in a sample of five observations under the variable name "Gender," we create a variable called "male" for dummy variable 1, as follows:

Gender	Male
Male	1
Male	1
Female	0
Female	0
Male	1

Note that only *one dummy variable* is required to represent *two categories* in a qualitative variable. Suppose another qualitative variable represents *three races* in a population. One single dummy variable would not be sufficient to represent all three types of race: White, Black, and Indian. Thus, we need to *create two dummy* variables. Let us assign the value of 1 for the race "White" in the first variable and the value of 1 for the race "Black" in the second variable. Hence, the race "Indian" receives the value of zero in either case. This is similar to the previous example where "female" received the value of zero. We can call these new variables by any names we wish, but logical and representative names would be "White" and "Black," respectively, because they receive the value of 1 for the associated race.

Race	White	Black
Indian	0	0
Black	0	1
Black	0	1
White	1	0

For the dummy variables you should have one less new column than the number of categories in your qualitative data.

What Happens with Too Many Dummies?

It is possible to create too many dummy variables. For example, creating one column where "male" is assigned the value of 1 and "female" is assigned the value of zero, and another dummy variable where the category "female" is assigned the value of 1 and "male" is assigned the value of 0. Similarly for the case of race having one dummy variable for each race will be result in having too many dummies, as shown in Table 8.1.

Table 8.1. Too Many Dummy Variables

Race	White	Black	Indian	Sum
Indian	0	0	1	1
Black	0	1	0	1
Black	0	1	0	1
White	1	0	0	1

Note that an additional column representing row totals is created under the name "sum." Also note that all the row values for "sum" are 1. The fact that the columns representing the dummy variables add up to one when *too many dummy* variables are created violates the assumptions of independence of X variables. This problem is called *perfect multicollinearity*. In the presence of perfect multicollinearity, the method of least squares cannot provide an answer. More sophisticated software gives a warning, and most likely would not report an output. Excel will present an output, but the first dummy variable in the data, that is, the one in the column furthest to the left among the over-specified dummy variables, will have the value of zero in the row depicting the name of the variable for all entries except for the *p* value, where the software will display "#Num!" This is the only indication that something is wrong. Do not be tempted to use other results from the output; they are wrong because there is no solution when there is *perfect multicollinearity*. Because the algorithm fails while attempting to find a solution, the exact output might not be the same under all cases. The more common and also more serious problem is when multicollinearity is not perfect but is high. The case of general multicollinearity is addressed in Chapter 9, *Regression Pitfalls*.

Advantages of Using Dummy Variables

Things should not be done one way or the other without consideration or expectation of an explanation. An observant student would propose the alternative of performing two separate regressions: one involving male data and another one involving female data. A similarly valid suggestion would be to perform three regression analyses for the three races instead of one with two dummy variables. The ideas are valid and would provide similar outcomes.

One advantage of using dummy variables instead of separate regression analyses, one for each category of qualitative variable, is that the *degrees of freedom* increase substantially by using dummy variables. Grunfeld (1958) collected current investment (I) and regressed it on beginning-of-year capital stock (C), and the value of outstanding shares at the beginning of the year (F), for 11 firms (General Motors, US Steel, General Electric, Chrysler, Atlantic Refining, IBM, Union Oil, Westinghouse, Goodyear, Diamond Match, and American Steel) from 1935 to 1954. Each of the two separate regressions of (I) on (C) and (F) would have 2 degrees of freedom for regression and 17 degrees of freedom for residual.[2] Therefore, the F statistics for each of the 11 regression analyses has 2 and 17 degrees of freedom and the t statistics for each of the two independent variables in each regression output has 17 degrees of freedom. Introduction of *10 dummy variables*, representing the 11 firms, results in 12 degrees of freedom for regression and 207 degrees of freedom for residuals. Therefore, the F statistics for this model has 12 and 207 degrees of freedom and each of the t tests for the 11 slopes plus the intercept has 187 degrees of freedom. Zellner (1962) pointed out that although the study of the 10 firms individually makes sense, there is important information that applies to all 10 firms and conducting 11 separate regression analyses will miss the extra information.[3] Thus, he created the *seemingly unrelated regression analyses*. The second advantage of using dummy variables is that excluding the information that could be captured by the seemingly unrelated regression is a mistake. The third advantage of using dummy variables is that testing many hypotheses *inflates the probability of type I error*, that is, the probability of rejecting the null hypothesis when it is actually true. When many hypotheses are tested, the researcher is under the false impression that he or she is testing a particular level of significance, while in reality it is much higher. A correction for such cases is called *Bonferroni correction*.[4]

Interpretation of Qualitative Independent Variables

Interpretation of dummy variables hinges on the fact that the contribution of the category that receives the value zero is represented by the intercept rather than the slope of the dummy variable, as is customary with other independent variables.

Let us regress I on C and F from Grunfeld data and, for the sake of simplicity, only consider two firms, General Electric and Westinghouse. Therefore, only one dummy variable is needed in model (8.1).

$$I = \beta_0 + \beta_1 C + \beta_2 F + \beta_3 D + \varepsilon, \qquad\qquad (8.1)$$

where D is the dummy variable representing firms General Electric and Westinghouse. We randomly choose Westinghouse to receive the value of 0 and General Electric to receive the value 1. For the rows representing Westinghouse, the value of D is zero and β_3 times zero results in a value of zero in the model, thus, the model reduces as follows:

$$I = \beta_0 + \beta_1 C + \beta_2 F + 0 + \varepsilon \qquad\qquad (8.2)$$

Therefore, the intercept β_0 represents the contribution of General Electric to the model. On the other hand, for Westinghouse the value of D is 1. Therefore, the model reduces to

$$I = \beta_0 + \beta_1 C + \beta_2 F + \beta_3 + \varepsilon \qquad\qquad (8.3)$$

The combined term $\beta_0 + \beta_3$ is actually the intercept of the model when dealing with Westinghouse. Except for a difference in intercept, the two models are identical. This clearly demonstrates that dummy variables only change the intercept of the regression model and leave the slopes of the remaining random variables, which are usually the main objective of the study, unchanged. Therefore, the impact or contribution of General Electric in the estimated equation is estimated by the original intercept, while that of Westinghouse is measured by the slope for the dummy variable. Nothing else is affected; therefore, the interpretation of the slopes of C and F remains unchanged.

Creating Dummy Variables in a Spreadsheet

One of the advantages of spreadsheet software is its data manipulation capabilities. Because spreadsheets are versatile, they offer several ways to create dummy variables. The choice of a particular way depends on the level of familiarity of the user with the software. The easiest, and most likely the method that everyone who is familiar with spreadsheet can do, is by sorting based on the qualitative data and assigning the value of zero or one as applicable, and copying it down to all the rows that are supposed to receive the assigned values. It is also possible to write a single nested "if" command to create all the necessary dummy variables. When using Grunfeld data, the "if" statement must be copied into 220 rows and 10 columns. One can automate the entire process, especially if there are repeated studies with similar modeling requirements using macros. Choose the method you are comfortable with. If you are not comfortable with any of those suggested here, you still have the option of entering the numbers zero and one individually by typing them. The example involving the male and female categorical data is a smaller example. There are only three 1's and two 0's; just type them in. For all small projects the best way is to simply enter the numbers. For larger databases, the logical thing to do is to take the time to utilize the software's capabilities. The investment in the learning and mastering the software will pay off in no time.

The use of qualitative independent variables is not limited to constant slope with different intercepts as in the above examples. With minor modification in stating the model, we can utilize dummy variables to estimate models that yield different slopes for different independent variables or some of them as needed. The subject, however, is beyond the scope of this text.

An Example

To account for the differences in firms in the Grunfeld data, investment is regressed on 10 dummy variables representing 10 of the firms. In this example, American Steel is represented by zero, so its effect is captured by the intercept (for the list of firms see above). The complete output is represented in Table 8.2. The model is

$$I = \beta_0 + \beta_1 C + \beta_2 F + \beta_3 D_1 + \beta_4 D_2 + \beta_5 D_3 + \beta_6 D_4 + \beta_7 D_1 + \beta_8 D_5 + \beta_9 D_6$$
$$+ \beta_9 D_7 + \beta_{10} D_8 + \beta_{11} D_9 + \beta_{12} D_{10} + \varepsilon \qquad (8.4)$$

Table 8.2. Regression of I on C, F, and 10 Dummy Variables

Regression statistics

Multiple R	0.9726639
R square	0.946075
Adjusted R Square	0.9429489
Standard error	50.299521
Observations	220

ANOVA

	df	SS	MS	F	Significance F
Regression	12	9,188,266.247	765,688.85	302.63881	4.7693E-124
Residual	207	523,718.6622	2530.0418		
Total	219	9,711,984.91			

	Coefficients	Standard error	t Stat	p-value	Lower 95%	Upper 95%
Intercept	-20.578198	11.2977936	-1.8214351	0.0699842	-42.85168947	1.6952936
Value	0.1101291	0.011299843	9.7460749	1.034E-18	0.087851587	0.1324067
Capital	0.3100334	0.016540477	18.743924	1.746E-46	0.277424051	0.3426428
D_1	-49.720869	48.2800578	-1.0298428	0.3042856	-144.9045389	45.462801
D_2	122.48294	25.9595257	4.7182271	4.374E-06	71.30398181	173.66189
D_3	-214.9912	25.46125786	-8.443856	5.362E-15	-265.1878214	-164.79457
D_4	-7.2309133	17.33822177	-0.4170505	0.6770737	-41.41305095	26.951224
D_5	-94.024318	17.16371479	-5.4780867	1.24E-07	-127.8624164	-60.186219
D_6	-2.5820021	16.37918574	-0.1576392	0.8748948	-34.87340943	29.709405
D_7	-45.966025	16.35747974	-2.8100921	0.0054278	-78.21463929	-13.717411
D_8	-36.968293	17.30915026	-2.1357659	0.0338731	-71.09311671	-2.8434698
D_9	-66.636345	16.37884196	-4.0684406	6.729E-05	-98.92707452	-34.345615
D_{10}	14.010167	15.94359656	0.8787332	0.3805648	-17.42248044	45.442814

The first thing to do is test the goodness of the model

H_0: Model is not good H_1: Model is good

The p value for F statistic is 4.77E–124, which is low enough to reject the null hypothesis. Therefore, at least one of the slopes is not zero, but each of the 12 parameters must be tested individually.

H_0: $\beta_i = 0$ for i from 1 ... 12

H_1:β_i ? 0   for i from 1 ... 12

where the "?" should be replaced with one of the signs, less than, greater than, or not equal, depending on the researcher's claim. For example, the expected signs of slopes β_1 and β_2 are both positive and the ">" should be used. Most researchers do not have a claim on the direction of the slopes of the dummy variables representing the firms. Therefore, the sign # should be used in the alternative hypotheses for dummy variables.

The estimated slope for American Steel represented by the intercept is –20.58; the corresponding p value is 0.07. This p value is not low enough to reject the null hypothesis. Dummy variables for US Steel (D_2), General Electric (D_3), Atlantic Refining (D_5), Union Oil (D_7), and Westinghouse (D_8) are significant while the rest are not. Of the dummy variables that are significant, that of US steel is positive, while the rest are negative. This means that "other things equal" the firms with positive coefficients for their dummy variables invested more than the base company, American Steel. The magnitudes of the firm effect are reflected in the coefficients of significant dummy variables. The only thing that differentiates the firms' investment behavior is the intercept of their estimated equations. These values indicate the amount of investment in a year for each company, if "value" (value of outstanding shares at the beginning of the year) and "capital" (beginning-of-year capital stock) were zero at the beginning of the year. The signs of the dummy variables indicate whether the corresponding firm's intercept is above, for positive coefficients, or below, for negative coefficients of the intercept for American Steel. The magnitude of each coefficient indicates how far the regression line of each firm is above or below the regression line for American Steel. The interpretations of slopes of the other two variables are exactly the same for all the firms.

The slope for "value" (value of outstanding shares at the beginning of the year) is 0.1101 and is significant at 1.034E–18 level, a very small probability of type I error. This means that for every $1 increase in value of outstanding shares at the beginning of the year, a company would invest 11 cents more in a given year.

The slope for "capital" (beginning-of-year capital stock) is 0.31 and is significant at the 1.746E–46 level, a very small probability of type I error. This indicates that a dollar more in capital stock at the beginning of the year would increase investment in that year by 31 cents. This example demonstrates the advantages of including dummy variables.

CHAPTER 9

Pitfalls of Regression Analysis

Forming Incorrect Hypotheses

Using regression analysis without a theory to guide it is a waste of time, if not outright dangerous. Let us start with a simple hypothetical example. Afterward, real data will be used to demonstrate the problem.

Assume we have collected the following data, without any theory whatsoever.

Y	X_1	X_2
11	5	12
13	12	10
17	16	6
19	19	4
21	39	1

Enter this data into Excel. Perform regression on the following two models:

$$Y_A = \beta_0 + \beta_1 X_1 + \varepsilon \tag{9.1}$$

$$Y_B = \beta_0 + \beta_2 X_2 + \varepsilon \tag{9.2}$$

The only reason for labeling Y with subscripts A and B is for reference. First regress Y on X_1. The result is shown in Table 9.1.

As usual, the hypothesis for goodness of fit is

H_0: Model is not good $\qquad$ H_1: Model is good

Table 9.1. Regression of Y on X_1

SUMMARY OUTPUT

Regression statistics

Multiple R	0.89224474
R Square	0.79610068
Adjusted R Square	0.72813425
Standard Error	2.1624271
Observations	5

ANOVA

	df	SS	MS	F	Significance F
Regression	1	54.77127211	54.7717271	11.7131441	0.041768073
Residual	3	14.02827289	4.67609096		
Total	4	68.8			

	Coefficients	Standard error	t Stat	p-value	Lower 95%	Upper 95%	Lower 95.0%	Upper 95.0%
Intercept	10.9200983	1.820775909	5.99749716	0.0092836	5.125576778	16.7146199	5.12557678	16.7146199
X_1	0.29010449	0.084765222	3.4224709	0.04176807	0.02034372	0.55998525	0.02034372	0.559865254

The p value of 0.04 is low enough to reject the null hypothesis that the model is not good. Therefore, the conclusion is that the model is good. The next step is to proceed with a test of hypothesis concerning the slope. In the lack of theory to determine the alternative hypothesis, it is tempting for a novice to look at the estimated slope, observe a positive value, and claim a direct relationship between Y_A and X_1. Note that this procedure is *WRONG*. We are presenting all these steps to reinforce that it is essential to learning processes and to build a habit of following a routine when performing a regression analysis.

$$H_0: \beta_1 = 0 \qquad H_1: \beta_1 > 0$$

The following is *WRONG*, due to lack of theory and use of a sample to determine the alternative hypothesis: The null hypothesis is rejected because p value for the slope of the equation is 0.04, which is low enough. It implies that if 100 similar regressions were performed on similar data and the null hypothesis is rejected, there are four chances of committing type I error. According to the $R^2 = 0.7961$, the model explains 79.61% of the variation in the dependent variable.

Next, regress Y on X_2 for the second model, namely:
As usual, the hypothesis for goodness of fit is

$$H_0: \text{Model is not good} \qquad H_1: \text{Model is good}$$

The p value of 0.00019 is low enough to reject the null hypothesis that the model is not good. The next step is to proceed with a test of hypothesis concerning the slope. Without a theory to determine the alternative hypothesis, it is tempting for a novice to look at the estimated slope, observe a negative value, and claim an inverse relationship between Y_B and X_2. Note that this procedure is *WRONG*. The result is shown in Table 9.2.

$$H_0: \beta_2 = 0 \qquad H_1: \beta_2 < 0$$

The following is *WRONG*, due to lack of theory and use of sample to determine the alternative hypothesis: the null hypothesis is rejected

Table 9.2. Regression of Y on X_2

SUMMARY OUTPUT

Regression statistics

Multiple R	0.997059336
R Square	0.99412732
Adjusted R Square	0.99216976
Standard Error	0.366987922
Observations	5

ANOVA

	df	SS	MS	F	Significance F
Regression	1	68.3959596	68.3959596	507.84	0.000191342
Residual	3	0.404040404	0.134680135		
Total	4	68.8			

	Coefficients	Standard error	t Stat	p-value	Lower 95%	Upper 95%	Lower 95.0%	Upper 95.0%
Intercept	22.33333333	0.317820863	70.2701928	6.35098E-06	21.3128855	23.3447811 6	21.3128855	23.3447811 6
X_2	−0.929292929	0.041237201	−22.5353056 3	0.000191342	−1.06052810 7	−0.798057751	−1.06052810 7	−0.798057751

because p value for the slope of the equation is 0.00019, which is low enough. It states that if 100,000 similar regressions were performed on similar data and the null hypotheses were rejected, there are 19 chances of committing a type I error, that is, 19 errors in 100,000 trials. According to $R^2 = 0.9941$, the model explains 99.41% of the variation in the dependent variable. Note the slope is designated by "2" to indicate that it belongs to variable X_2.

In summary, the novice concludes that when Y is regressed on X_1, there is a direct relationship between the two, and when the same Y is regressed on X_2, there is an inverse relationship between the two.

Multicollinearity

In the previous section, two examples were used to discuss the problems of using data to determine a hypothesis and then testing it. In this section, the same two examples are combined to show another problem, commonly known as multicollinearity. When independent variables in multiple regressions are not independent, there exists *multicollinearity*.

In the previous section, regressions of dependent variable Y on each of the independent variables X_1 and X_2, respectively, provided fairly good estimates, although meaningless. The relationship between Y and X_1 is direct, and the one between Y and X_2 is inverse. Now regress Y on both X_1 and X_2, showing the dependent variable as Y_C to distinguish it from the previous two regressions. The proposed model is written as

$$Y_C = \beta_0 + \beta_1 X_1 + \beta_2 X_2 + \varepsilon \qquad (9.3)$$

The results are shown in Table 9.3.

All the metrics show improvements. The p value for F statistics for testing goodness of the model is 0.00046, which is low enough to reject the null hypothesis:

H_0: Model is not good H_1: Model is good

Table 9.3. Regression of Y on both X_1 and X_2

SUMMARY OUTPUT

Regression statistics	
Multiple R	0.999767757
R Square	0.999535568
Adjusted R Square	0.999071136
Standard Error	0.12639805
Observations	5

ANOVA

	df	SS	MS	F	Significance F
Regression	2	68.76804709	34.38402354	2152.168421	0.000466432
Residual	2	0.031952911	0.015976456		
Total	4	68.8			

	Coefficients	Standard error	t Stat	p-value	Lower 95%	Upper 95%	Lower 95.0%	Upper 95.0%
Intercept	24.55265924	0.474722277	51.93886656	0.000370488	22.51869945	26.58661903	22.51869945	26.58661903
X_1	-0.062224091	0.012893661	-4.825944415	0.040356033	-0.117701036	-0.006747146	-0.117701036	-0.006747146
X_2	-1.093966786	0.036960465	-29.59829613	0.001139525	-1.252994831	-0.934938741	-1.252994831	-0.934938741

The two separate hypotheses for coefficients of X_1 and X_2 are also repeated below:

$$H_0: \beta_1 = 0 \qquad H_1: \beta_1 > 0$$
$$H_0: \beta_2 = 0 \qquad H_1: \beta_2 < 0$$

The corresponding p values for X_1 and X_2 are 0.04 and 0.001, respectively. Because both are small enough one would be tempted to reject each hypothesis in favor of its respective alternative hypothesis. An observant reader, however, would recognize that doing so would result in *type III error* in the case of the slope of X_1 variable; because the alternative hypothesis is positive while the slope has a negative sign.

Once again, this demonstrates the problem of using data to form hypotheses. A hypothesis has to be formed based on theory. If there is no theory to guide you, or there is no prior knowledge of a variable and its relationship to the dependent variable under consideration, then it would be reasonable to obtain a sample to find an estimate of possible relationship between the dependent and independent variable. Use the obtained sample to form a claim and write an alternative hypothesis, then *collect another sample* to test that hypothesis. This simple example demonstrates the potential problem of letting data determine the alternative hypothesis, which in this case results in having two contradictory alternative hypotheses.

The reason for getting contradictory slope signs evidenced in the above example is correlation between independent variables. This violates the requirement of independence of exogenous variables, which is discussed in Appendix. This problem is very common in economics because most, if not all, economic variables are interrelated. The correlation coefficient between X_1 and X_2 is -0.923. The example is designed to highlight the problem, however, in real data such high degree of correlation among economic variables exists more often than one might expect. Even if the correlation coefficient was not as high, regression coefficients would be different for the case of regressing Y on each of the X values than corresponding coefficients when Y is regressed on both X_1 and X_2. It is vital to use economic theory in forming a regression model and to specify alternative hypotheses for all coefficients based on their anticipated theoretical values. Even then the problem is a serious one and is called *spurious* correlation in the literature.[1] As in the above example, an incorrect sign shows the presence of spurious relationship. In less extreme cases, the magnitudes of coefficients will change but not necessarily their signs. This makes it more difficult to suspect spurious correlation. In such cases the researcher's knowledge of the subject will be invaluable. Any outcome that does not conform to the theory or is unexpected should be a cause for alarm. For example, if income elasticity of a good which is believed to be a necessity, such as potatoes, happens to be greater than 1 in a sample, which would indicate it is a luxury, it is a cause for an alarm.

In *cross-sectional analysis*, spurious regression refers to the situation where two exogenous variables are correlated with each other, or each is correlated with a third variable, which is often not in the model. In *time series analysis*, the term refers to a situation where one trending variable is regressed on another trending variable. In the latter case, data must be *de-trended*. In cross-sectional cases, the preferred solution is to identify the variable that is correlated with both variables in the model, the "cause variable," and use it in the model instead of the two "effect" variables. If this is not possible, then one of the variables must be excluded from the model. The full coverage of spurious regression is beyond the scope of this book.

Multicollinearity does not affect the overall estimation of the dependent variable. It affects statistical significance of individual exogenous variables. The *t* test fails to reject the null hypothesis while the slope of the variable is actually different than zero, thus resulting in a *type II error*. The issue is that when assessing the contribution of a coefficient, the assumption is to "keep other things equal," which is not possible because of multicollinearity. Some argue that it is perfectly acceptable to use regression models to estimate the dependent variable in the presence of multicollinearity, as long as no specific remark is made about the contribution or even the significance of individual independent variables.

Detecting Multicollinearity in Cross-Sectional Data

Detecting multicollinearity is easy. All that is needed is to determine whether exogenous variables are independent from each other. Although it is possible to calculate the pair-wise correlation coefficient for all variables, it is easier and better to use the *variance inflation factor (VIF)*. MS Excel does not calculate VIF but the formula is fairly straightforward but tedious.

$$VIF = 1/(1 - R_j^2), \qquad (9.4)$$

where R_j^2 is the R^2 for regressing each independent variable on all the other independent variables. Therefore, there would be one *VIF* for each slope. If *VIF* is greater than 10, there is multicollinearity. The denominator, that is $(1 - R_j^2)$, is called *tolerance level*. If the tolerance level is less than 0.1, there is multicollinearity. Another recommendation is that if the average of the *VIF*s is greater than 1, there exists some multicollinearity.

Another easy way to detect multicollinearity is the presence of significant F statistics together with no or few significant t statistic. A significant F statistic indicates that the model is good and able to estimate the dependent variable well. Insignificant t statistics indicate that the slopes of independent variables are not different than zero. Therefore, when combined, the independent variables are effective in estimating the dependent variable, but individually they are meaningless. This is the essence of lack of independence among the exogenous variables.

The *VIF* formula is part of the formula for the variance of the slope coefficient. A large *VIF* means large variance, and thus, a small t statistic for the slope, which makes it more difficult to reject the null hypothesis of zero slope. The other component of the variance of the slope is proportionate to the variance of the independent variable, which is another cause of lack of significance of a slope while the F statistics indicates a good overall fit.

Issues with Independent Variables

When the dependent variable Y is written as a function of independent variable X in regression, it means that a change in a particular X will change Y by the magnitude of the coefficient for X, that is β, other things equal. The implication is that if X does not change, Y should not change either. Although the relationship can be real and theoretically correct, the data must be suitable to allow testing the pertinent hypotheses. In economics, numerous independent variables are included in the model to account for theoretical requirements and also to control for variables that are required to be constant by economics theory. Furthermore, the data are collected from existing sources and are not experimental, which means exogenous variables are not "controlled" in the sense of natural science experiments. It is important to make sure that each independent variable X contains sufficient variability and be as spread as possible.[2] Otherwise, its contribution to changes in Y cannot be measured. The Y values usually have sufficient variability due to contributions from all independent variables. An observant student would notice that the above statement means that the variance for each X should be as large as possible. This is contrary to the general recommendation to reduce variance. The difference is that here we are talking about independent variables or exogenous variables instead of the dependent or endogenous variable. There is no theoretical need for small variance for an independent variable; on the contrary, the larger the variances of the independent variables the better.

The majority of economic data are either short or must be shortened to make them meaningful. This is true of both time series and cross-sectional data. Most time series economic data are from post-World War II (WWII). Reliable data do not exist for all variables for periods prior to WWII. When one variable has a missing value, the remaining variables must be excluded from the analysis, which shortens all the available data. The other reason is that major events such as WWII change the nature of economic activities. In the 1950s and even 1960s, when studies needed sufficient data, the data from 1939 to 1945 were excluded because of this problem. Furthermore, economic relationships change over time. Demands for many products are different in 2012 than they were in 1950, and therefore, the variables that need to be included in the model are different. For example, in a study of demand for live music, a type of entertainment, it is necessary to include dummy variables for new products such as iPod and laptop computers, which were not available until recently. The result is shortening the length of the data series. Cross-sectional data are shortened due to changes in the units of observation. Since 1989 there have been numerous new countries for which there are no prior data. Another example is the change in income-earning structure of family.

Linearity

The procedures that have been covered in this text are designed for linear models. Linear models are much broader than the topics covered in this book. However, there are times when linear models are not reasonable. In fact, most of economics relationships are not linear. Assuming linearity simplifies modeling. Even in the simplest of economic topics such as demand function, linearity is assumed and not necessary except for simplicity. Nonlinear demand functions can be very interesting to study. For example, a constant elasticity demand function is nonlinear. One of the shortcomings of linear demand functions is that the location on the line determines whether a good is elastic or inelastic. Therefore, as the price changes, the elasticity of demand changes along the linear demand function. Figure 9.1 depicts a nonlinear demand function.

Nonlinear statistical analysis is more complex than linear analysis. Sometimes it is possible to convert a nonlinear function to a linear function using simple mathematical transformations to obtain linear models, consequently, reducing the magnitude of complexity of computation and analysis by many folds. Sometimes, it is not even meaningful to have a linear model, as is the case with production function.

Definition 9.1

A *production function* indicates the maximum potential output that can be produced with a set of input at a given level of technology.

In practice, determining actual output is very important. Note that here technology is given and thus must be controlled in a regression

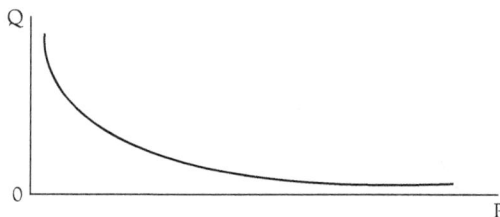

Figure 9.1. A nonlinear demand function.

model and somehow represented as an independent variable. The function is nonlinear due to the law of diminishing marginal productivity.

Definition 9.2

The *marginal product* of an input is the amount of increase in output as a result of one unit increase in that input, other things being equal.

Definition 9.3 The law of diminishing marginal productivity

Continuous increase of one input while other inputs are kept constant will eventually lead to a reduction of marginal product of that input. At extreme and under certain conditions total product might also decline. A typical production function is depicted in Figure 9.2.

A commonly used production function is the Cobb–Douglas production function.

$$Q = TL^a K^{(1-a)} U, \tag{9.5}$$

where Q is output, L is labor, K is capital, T is technology, α is the share of labor productivity in output, and U is statistical error. Taking the natural logarithm of the model converts the nonlinear model of equation (9.5) into linear function shown in equation (9.6).

$$\mathrm{Ln}\,Q = \mathrm{Ln}\,T + a\,\mathrm{Ln}L + (1 - a)\,\mathrm{Ln}K + \mathrm{Ln}\,U. \tag{9.6}$$

This is the same as typical regression with the exception that all variables are in the natural logarithm forms.

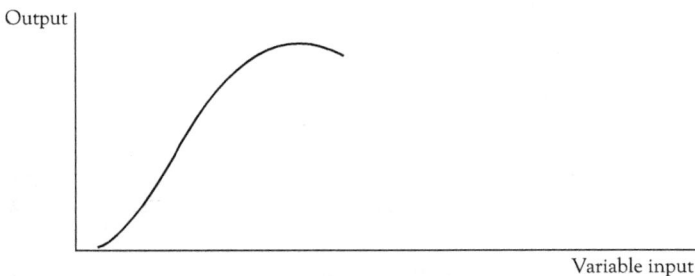

Figure 9.2. A production function.

$$Y = \beta_0 + \beta_1 L + \beta_2 K + \varepsilon. \tag{9.7}$$

Before performing the regression analysis on the data for model (9.7) take the natural logarithm of all the variables. To obtain natural logarithm using Excel use the following command:

$$= Ln(X), \tag{9.8}$$

where "X" can be a number or the address of a single cell in the spreadsheet. An important point to remember is that the entire function must be of multiplicative form, that is, the only permissible operator is the *product* sign. Therefore, the following cannot be converted into a linear form because one of the variables, namely the error, has an additive relation.

$$Q = TL^a K^{(1-a)} + U. \tag{9.9}$$

Before long you would run into a model of the form

$$LnQ = LnT + a\, LnL + (1-a)\, LnK + U. \tag{9.10}$$

In model (9.10) all the variables are in natural logs except the error term. Models containing variables in levels as well as in natural log are also used in the literature. The interpretation of coefficients and the changes in the dependent variable in such models are very difficult. In such cases, the analysis is limited to statements such as "if X changes by one unit, Y will change by β units." Other economic interpretations of such models are next to impossible. However, when all the variables are in the level units, then the changes are also in level units. On the other hand, when all the variables are in natural logs, then the slopes represent *rates of change*. Rates of change have many applications and uses in economics. May be the most familiar is the price elasticity of demand. The price elasticity of demand represents the percentage change in quantity for a percentage change in price. When the natural log of quantity is regressed on the natural log of price, the coefficient for the price is the slope. Therefore, it is recommended to either use multiplicative models that can be converted into linear models or use additive models to begin with.

The above statement does not mean that you should never modify data; it only clarifies the point that the economic interpretations of such mixed models are more difficult. In more advance courses you will learn that it might be necessary to transfer some of the original variables to make them stationary or approach normal distribution, among other reasons.

Other types of transformations are conceivable as well. In fact, taking the square root or raising variables to powers of two or larger is a common practice. The former is usually beneficial in transforming non-normal data, while the latter might be necessary due to model assumption or in an attempt to capture decreasing or increasing marginal effects. A classic example is Kuznets' income inequality in which he claims that "as a country begins its development, income inequality worsens at first then improves." The claim is depicted in Figure 9.3.

In order to test this hypothesis we need to find an appropriate variable to represent economic development and income inequality. A common variable for economic is the growth domestic product (GDP) and an acceptable variable for income inequality is the Gini coefficient.[3]

$$\text{Gini} = \beta_0 + \beta_1 GDP + \beta_2 GDP^2 + \beta X + \varepsilon, \qquad (9.11)$$

where βX represents all the other pertinent factors. A major issue concerning models of this form is the inability to keep other things equal. Therefore, it is not possible, for example, to talk about the impact of one unit change in GDP on Gini or one unit change in GDP^2. A one unit change in GDP affects Gini in two ways, once through β_1 because of

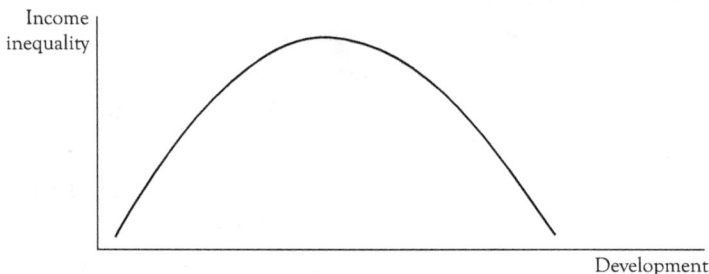

Figure 9.3. Income inequality and development.

GDP and simultaneously through β_2 because of GDP^2. There is another factor that involves calculus. All the slopes are actually partial derivatives of the dependent variable with respect to different independent variables. Since until now we were only dealing with independent variables that had power of one (1), the results of partial derivatives were the slopes. However, now there is as follows:

$$\text{Effect of one unit change in } GDP \text{ on Gini} = \beta_1 + \beta_2 GDP \qquad (9.12)$$

APPENDIX A

Regression Assumptions

Need for Assumptions

This appendix explains necessary assumptions about *error term* in regression analysis and addresses other important issues that are essential for correctness of the results. In order to be able to use theories in statistics to perform statistical analysis, it is necessary to adhere to the rules that govern those theories or are prerequisite for their validity. The assumptions that are listed and discussed below are necessary for validity of inferences. The proofs of their contribution to the theory and the validity of the conclusion are left for specialty textbooks. However, before addressing the assumptions, let us change the notation of equation (1.9)

$$C = \beta_0 + \beta_1 Y + \varepsilon \qquad (A.1)$$

to a more generic equation where the dependent variable is designated by the letter Y, and the independent variable is designated by the letter X. In this formulation, the letters "Y" and "X" are merely placeholders for the dependent and independent variables, respectively.

$$Y = \beta_0 + \beta_1 X + \varepsilon \qquad (A.2)$$

Assumption A.1

In a regression model, the relationship between dependent and independent variables is linear in parameters. The model in (A.2) is an example where the linearity assumption is met.

The linearity assumption is different from the definition of a line, which requires the powers of X and Y to be equal to one. The following model still meets linearity assumption and is considered a linear regression model, while it does not represent a line in geometric sense.

$$Y = \beta_0 + \beta_1 X^2 + \varepsilon. \qquad (A.3)$$

Functions with powers of 2 are known in mathematics as quadratic functions. To see how model (A.2) meets the linearity assumption, we define a new variable $Z = X^2$.

$$Y = \beta_0 + \beta_1 Z + \varepsilon. \qquad (A.4)$$

Model (A.4) is identical to model (A.2).

Assumption A.2

Data are obtained by random sampling. This means that observation must be *chosen at random*. It does not mean that the observations must be random.

Assumption A.3

Conditional expected value of the error given the independent variable is zero. The key concept here is conditionality. It means that given a particular value of an independent variable, the average of the error term is zero. The conditional relationship between the error term and the independent variable is extended to a conditional relationship between the dependent variable and independent variable. In other words, for a given value of X there exist many potential values of Y, one of which is observed in the sampling pair (X, Y).

The consequence of this assumption is that conditional expected value of the dependent variable given the independent variable is

$$E(Y|X) = \beta_0 + \beta_1 X. \qquad (A.5)$$

The estimated "expected" value of Y from regression is

$$\hat{Y} = \widehat{\beta_0} + \widehat{\beta_1} X. \qquad (A.6)$$

Assumption A.4

The conditional variance of the error given the independent variable is constant σ^2.

The consequence of this assumption is that conditional variance of the dependent variable given the independent variable is

$$\text{Var}(Y|X) = \sigma^2 \qquad (A.7)$$

Another common way of stating the above assumptions is to say that the dependent variable is a linear function of the independent variable where the random errors of the model have independent and identical distribution with mean zero and a constant variance, which is depicted by σ^2.

Statements in assumptions A.3 and A.4 involve the word *conditional*. This is very important because without requiring the presence of an independent variable the only statement we can make about the dependent variable would be based on observations of the dependent variable. This reduces to statements based on the average of dependent variables.

For example, in the model about consumption

$$\text{Consumption} = \beta_0 + \beta_1 \text{ Income} + \varepsilon \qquad (A.8)$$

Assumptions A.3 and A.4 state that for every value of income there are many possible values of consumption. The actual observed value of consumption is just one of those possibilities chosen at random. The expected value of a particular consumption value is given by E (consumption) = $\beta_0 + \beta_1$ Income. If the *endogenous variable* (consumption) did not depend on the exogenous variable (income) the best estimate for the endogenous variable would have been its mean (average consumption). However, because the dependent variable is *conditional* on the exogenous variable, the best linear estimate is the regression line.

Parameters β_0 and β_1 are usually *unknown* and must be estimated by regression analysis, which, according to Chapter 3, are estimates with some desirable properties. One such desirable property is the fact that the *mean of squared error* is the *smallest* possible mean squared error when the method of least squares is used. More detail on this issue is discussed below.

Assumption A.5

It is essential that the *error term* be *random*. The error term is not the same as error. The definition of error is repeated below:

Error is the proportion of variation among the values of the dependent variable that cannot be explained.

In Chapter 2, several distinct but related "errors" or error-based concepts were discussed. All of these "errors" are *statistics*, which indicates that they are obtained from samples. Some of the "errors" have specific names based on their origin and use, such as variance or sum of squared residuals. Many errors have several names or could be called by other names, meaningfully, such as *standard deviation*, which is also the *average error*. The standard deviation represents the average error of using the mean to predict or estimate observations. All of these "errors" are actually random variables in their strict statistical terms; they are obtained from a sample and are known by virtue of computation of observed values.

The *error term* in a regression model, however, is a property of the model. The error term is not observable and does not come from a sample, although it can be estimated using sample data and regression analysis. The estimated values are statistics, which were mentioned in the previous paragraph

It is of vital importance not to mistake any of the concepts of errors with *errors in measurement*. Errors in measurement refer to incorrectly measured or recorded values of dependent or independent variables. In some social science disciplines, the measurement error is called *validity*. It is also important to note that the only requirement for randomness is for the error term. There is no assumption or requirement that independent variables be random. In fact, if independent variable were random it would introduce the possibility that it is not independent from the error term, which has some serious consequences. There is no randomness requirement for the dependent variable either. However, the dependent variable, by virtue of being a linear function of the error term and the randomness requirement of the error term, also becomes a random variable. The dependent variable has the same distribution function as the error term with the following properties:

$$E(Y|X) = \beta_0 + \beta_1 X \qquad\qquad (A.9)$$

$$\text{Var}(Y|X) = \sigma^2 \qquad\qquad (A.10)$$

We still have not addressed the distribution function of the error term because none of the discussion so far required any particular distributional properties. Nevertheless, like any random variable, the error term has a distribution function that governs and determines the outcome and the value of what is observed. The distribution function could be, and in fact for most real-life events, is unknown. The discussion about the properties and consequences of distribution function of the error term is limited to the ones that are either known or can be approximated by a known distribution function.

The main need for assumptions about error term is to allow using mathematical statistics to derive desirable properties for the estimates of parameters of the model. If the estimates of parameters lack these desirable properties, they are useless for all practical purposes.

Desirable Properties of Estimators

A statistics, as defined above, provides an estimate for population parameter. A more correct term is a *point estimate*. There are three properties that make a point estimate useful. An estimate must be *unbiased, consistent,* and *efficient*. Estimators that can assure these properties are considered desirable estimators. These terms have precise definitions with assumptions and conditions that are the subject of mathematical statistics. Here, however, strict definitions and necessary assumptions are relaxed for the sake of simplicity. All *estimates are variables* because they are obtained from sample data; they are *statistics*. They are used to make inferences about their corresponding *parameters*, which are characteristics of population and are always *constant*.

Definition A.1

Unbiasedness

An estimate is an unbiased estimate of the corresponding population parameter if its expected value is equal to the population parameter.

For example, the sample mean is an unbiased estimate of population mean.

$$E(\hat{\mu}) = \mu \qquad \text{(A.11)}$$

The expected value of an unbiased slope of a regression equation (a statistics) is equal to the slope of regression model (a parameter).

$$E(\hat{\beta}_1) = \beta_1 \qquad \text{(A.12)}$$

Theorem A.1

If error terms in a regression model are independent, then the estimated slopes and estimated intercept are unbiased estimates of slopes and intercept of the regression model.

$$E(\hat{\beta}_1) = \beta_1 \qquad \text{(A.13)}$$

$$E(\hat{\beta}_0) = \beta_0 \qquad \text{(A.14)}$$

Failing to comply with the assumption would make the theorem inapplicable.

Definition A.2

Consistency

An estimator is a consistent estimator if its variance becomes smaller as sample size increases.

For example, the sample mean is a consistent estimator of the population mean. Because variance is a measure of error, as sample size increases, the error of sample mean in estimating population mean becomes smaller.

Theorem A.2

If the error terms of regression model are independent, then the estimated slopes and the estimated intercept are consistent estimates of the slopes and the intercept of the regression model.

Therefore, the error of estimated slope becomes smaller as sample size increases. Eventually, as sample size approaches infinity, estimates of slopes approach slopes of the population.

Definition A.3

Efficiency

Efficiency is a comparative property. An estimate is said to be more efficient than another estimate if its variance is smaller than the variance of the other estimator. Sample mean is the most efficient estimator of population mean among all its unbiased estimates. This does not preclude the existence of a more efficient but biased estimate for population mean.

Theorem A.3

If error terms in a regression model are independent, then among all the unbiased estimators of the intercept and slopes of the model, the least square estimates are the most efficient estimates.

In other words, slopes and intercept estimates obtained by the method of least square have the smallest variances among all the unbiased linear estimators of the slopes and intercept. These estimates are called the *best linear unbiased estimators (BLUEs)*. The text provided two examples for estimating consumption. One was the average consumption and the other was the regression estimates. The estimates obtained using regression based on the slope and intercept from the least square methods have smaller variances than the estimates based on the average consumption.

The theorems that are listed above are broader than the form in which they are presented here. An important point to remember is that no specific distribution function has been required for the error term.

Often, econometric books submit the above into a single theorem called Gauss–Markov theorem. If the assumptions are not valid or if the rules are broken, then the premises of the Gauss–Markov theorem would not hold either.

Violations of Regression Assumptions

The presumption in this book is that the assumptions for regression analysis are met. In this section, a brief discussion of consequences of violating some of the regression assumptions is presented. Detecting and remedying the problems caused by such violations are beyond the scope of this book, as they require more sophisticated and specialized software than Excel.

Assumption A.6

The distribution function of the error term is *normal* with mean μ and variance σ^2. By far, the assumption of normality is the most important assumption because without it the goodness of fit of regression cannot be tested using the F statistics. It is also impossible to test hypotheses about regression coefficients using t statistics; the same is true about building confidence intervals for the model or for the parameters.

The consequence of normality assumption for error term is that the dependent variable, Y, has a normal distribution as well. The variance of the dependent variable (Y) values is the same as the error term, namely σ^2. However, the expected value of the dependent variable is

$$E(Y) = \beta_0 + \beta_1 X \qquad (A.15)$$

Therefore, there are as many expected values for Y as there are pairs of data. For each value of independent variable (X) there are infinite possible values of Y, all of which come from the governing normal distribution having an average equal to $\beta_0 + \beta_1 X$ and a variance equal to σ^2. Figure A.1 below depicts the relationship. The line that connects the expected values of the Ys, that is, the conditional means

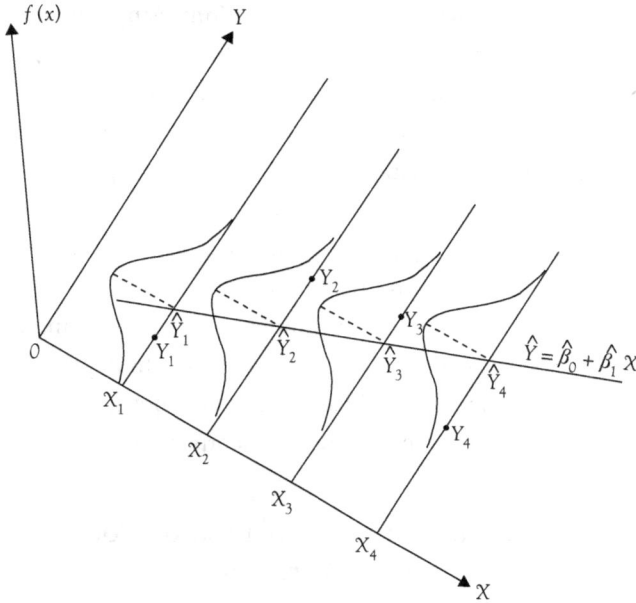

Figure A.1. Conditional distribution of Y with normality assumption.

of each *Y* distribution corresponding to each of the *X* values, is the regression line.

Note in Figure A.1 that the spread or the variances of the *Y* values depicted by the four normal distribution functions are the same.

Testing and Remedy for Violations of Normality

One of the ways to check for violations of normality is through tests for skewness and kurtosis.[1] Both statistics can be calculated in Excel using *Skew* and *Kurt* functions. The skewness and kurtosis tests refer to deviation from *symmetry* and *pointedness*, as compared to normal distribution, respectively. *Kolmogorov–Smirnov* and *Shapiro–Wilk* tests are two nonparametric tests of normality. These statistics, although simple to calculate, are not available in Excel, but online help with instructions to perform these tests are available. Another alternative is to consult one of the dedicated statistical analysis software.

Some textbooks suggest graphical inspections using numerous available graphs. This practice is a waste of time. The experts need to be able to test the hypothesis of normality precisely; a novice lacks the experience to differentiate between a normal and non-normal distribution by viewing graphs, except when the violation is flagrant. In cases of possible violation of this and other regression analysis assumptions, the reader is advised to move to the next level and gain the necessary knowledge to test the possibility of violation and to solve the problem.

Under some conditions, and in simple cases, transforming data in a monotonic way such as taking natural logarithm might improve the situation. If normality cannot be established, tests of hypotheses are meaningless and nonparametric methods must be incorporated.

Heteroscedasticity or Violation of Constant Error Variance

The assumption that variances of error term are constant, which is also known as *homoscedasticity*, is an important one. The problem of *heteroscedasticity*, or lack of equal variances, is more prevalent in cross-section data than in time series data. The consequence of *heteroscedasticity* is that regression coefficients are *inefficient*, although they are still *unbiased and consistent*. As indicated earlier in this appendix, efficiency refers to the size of the variance of estimates of a parameter, in this case the slopes, compared to variances of other estimation methods. When error variance is large, slope variance is also large, which in turn makes t statistics small. Consequently, the statistics might not be large enough to be statistically significant and results in failing to reject the null hypothesis, erroneously—a *type II error*.

When heteroscedasticity is present, it can be shown that *variances of slopes* of regression are *biased*, invalidating the tests of hypotheses about parameters and confidence intervals.

Because variances are different in the presence of heteroscedasticity, the spread of the values for dependent variable are not the same. Compare Figure A.2 with Figure A.1 and notice the difference between the spreads of the normal distribution functions.

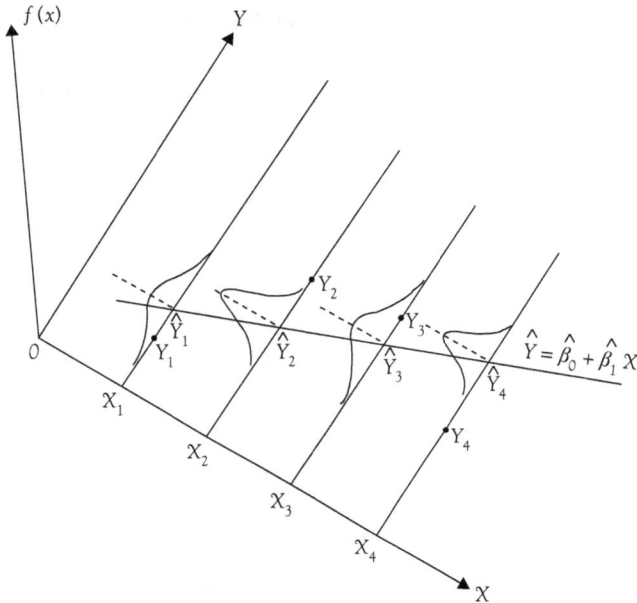

Figure A.2. Conditional distribution of Y under heteroscedasticity.

Testing and Remedy for Heteroscedasticity

The most common tests for heteroscedasticity are *Breusch–Pagan test* and *White test*. Neither is available in Excel and they require dedicated statistical software. The best solution to deal with heteroscedasticity is to use *White's robust* procedure. This method is beyond the scope of this book.

Serial Correlation

Serial correlation is the violation of independence of error terms. Serial correlation is more common in time series data than in cross-section data, especially for adjacent time periods or in the case of cyclical data when the data points represent similar cycles, for example, the fourth quarter. In majority of time series data, the problem of positive or direct correlation is more common than the negative or inverse correlation. One possible source of serial correlation is the inclusion of a lagged dependent variable as an exogenous variable.

Serial correlation *does not affect unbiasedness* or *consistency* of regression coefficients. However, *efficiency is affected*. Variances of slopes are be biased downward, indicating that they are be smaller than the true variances, especially when there is positive serial correlation. Therefore, *t* statistics will be larger than usual and the null hypothesis of zero slope would be rejected more often than they should be, resulting in *type I errors*. Serial correlation does not affect the coefficient of determination or R^2.

The serial correlation in time series is also called *autocorrelation*. Interestingly, the presence of autocorrelation can be used to provide better prediction for future values. In fact, some of the foundations of time series analysis are based on the use of serial correlation.

Testing and Remedy for Serial Correlation

There are numerous tests for different types of serial correlation, from a simple *t* test in the case of autoregressive of order one to a classical test of *Durbin–Watson* to *Breusch–Godfrey* test for higher order autoregressive cases. An autoregressive model is a time series model where the value of the endogenous variable is a weighted average of its own previous values.

The remedies also differ based on the situation. Some possible solutions when serial correlation exists include first *differencing, generalized least squares, and feasible generalized least square (FGLS)*.

Glossary of Terms

Adjusted *R* Squared provides a correction to *R* square based on the number of independent variables in the model.

Better is a concept that refers to the smallness of the average squared error.

Ceteris paribus is (Latin) for "other things being equal."

Coefficients are parameters in a regression and they are sometimes called slopes.

Coefficient of determination is another name for *R* squared. It shows the percentage of variation in the dependent variable that is explained by the regression line, over and above the average of the dependent variable.

Conditional expected value represents a change in the exogenous variable (Y) as a result of a unit change in a particular X.

Cross-sectional analysis is any analysis using cross-sectional data

Dependent variable is commonly used in economics literature, other names such as endogenous variable, Y variable, response variable, or even output are often used as well.

Descriptive statistics provides simple yet powerful insight to economic agents and enable them to make more informed decisions.

Elasticity is a measure of the responsiveness of one variable to changes in another.

Error is the difference between an observed value and its expected value.

Errors in measurement refer to incorrectly measuring or recording the values of dependent or independent variables.

Explained variance is sometimes called mean square regression.

***F* statistics** is the measure that verifies whether the explained portion of the dependent variable exceeds the unexplained portion.

Independent variable is commonly used in economics literature, other names such as exogenous variable, X variable, regressor, input, factor, or predictor variable.

Individual error is the difference between an observed value and its expected value.

Inferential statistics allows the economist and political leaders to test hypotheses about economic condition.

Marginal product of an input is the amount of increase in output as a result of one unit increase in that input, other things being equal.

Marginal propensity to consume represents the amount one would consume if one is given an extra dollar.

Mean is one of numerous statistical measures at the disposal of modern economists. It is an arithmetic average.

Mean squared error is the same as variance of a regression model.

Median is a value that divides observations into two equal halves. It is the midpoint among a group of numbers ranked in order.

Method of least squares is another term for regression analysis.

A **model** is a simple representation of something real in life.

Multicollinearity is the high correlation between two variables.

Omitted variable bias is when relevant variables have been excluded.

Ordinary least squares (OLS) is a regression analysis model with one variable.

P **value** represents the probability of type I error for inference about a coefficient.

A **parameter** is a characteristic of a population that is of interest; it is constant and usually unknown.

Price elasticity of demand is the percentage change in quantity demanded divided by the percentage change in price.

Probability is the likelihood that something will happen, expressed in the form of a ratio.

Production function indicates the maximum potential output that can be produced with a set of input at a given level of technology.

Random error, symbolized by the Greek letter Epsilon (ε), is also known as random component or error term.

Regression analysis is the effect of one or more factors measured to determine another factor. The first group is also known as *explanatory variables*, while the latter is known as *endogenous variables*.

Residual sum of squares is the amount of variation in the dependent variable that is unexplained by either the mean of dependent variable or regression line.

Sample statistics are used to make an informed decision about the average price of a product.

Sample variance consists of squared values of individual errors divided by degrees of freedom.

Scatter plot is a graph customarily used in presenting data from a regression analysis model.

Shifters in economics refer to variables that are assumed to remain constant for the sake of identifying the impact of the explanatory variables on the response variables.

Spurious correlation is when two variables are incorrectly inferred to be related to each other.

Standard error is the average error of a regression model.

A **statistic** is a numerical value calculated from a sample that is variable and known.

Statistical inference is used when restrict and formal statistical methods are performed.

Stochastic means that a model is probabilistic in nature and would result in varying results reflecting the random nature of the model.

Sum of squared errors (*SSE*) is always non-negative.

A **testable hypothesis** is a claim about a relationship among two or more variables.

Time series analysis is any analysis using time series data.

Tolerance level is a measure based on the desired level of risk and the probability of the outcome of an event of interest.

Type I error is rejecting the null hypothesis even though it is true.

Type II error is failure to reject a false hypothesis.

Type III error is rejecting a null hypothesis in favor of an alternative hypothesis with the wrong sign.

Unexplained variance is sometimes called mean square error.

Validity is the lack of measurement error.

Notes

Chapter 1

1. Naghshpour (2012).

Chapter 2

1. Naghshpour (2012).
2. Draper and Smith (1998).

Chapter 5

1. Naghshpour (2012).

Chapter 8

1. Naghshpour (2012).
2. Grunfeld (1958).
3. Zellner (1962).
4. Nickerson (1994).

Chapter 9

1. Granger and Newbold (1974).
2. Goldberger (1991).
3. Kuznets (1995).

Appendix

1. Naghshpour (2012).

References

Beer Institute. (2011). *Brewers Almance 2001: Per capita consumption of beer by state 1994–2010.* http://www.beerinstitute.org/statistics.asp?bid=200

Bureau of Economic Analysis. (2012a). *National income and product account tables: Personal consumption expenditures by major type of product.* http://www.bea.gov/iTable/iTable.cfm?ReqID=9&step=1

Bureau of Economic Analysis. (2012b). *GDP and personal income: Personal income summary.* http://www.bea.gov/iTable/iTable.cfm?ReqID=70&step=1

Bureau of Labor Statistics. (2011). *Consumer price index-average price data: Malt beverage, all types, all sizes, any origin, per 16 oz.* http://data.bls.gov/pdq/SurveyOutputServlet

Draper, N. R., & Smith, H. (1998). *Applied regression analysis* (3rd ed.). New York: Wiley-Interscience.

Goldberger, A. S. 1991. *A Course in Econometrics.* Harvard University Press.

Granger, C. W. J., & Newbold, P. (1974). Spurious regressions in econometrics. *Journal of Econometrics 2,* 111–120.

Grunfeld, Y. (1958). *The determinants of corporate investment.* Unpublished Ph.D. Dissertation, University of Chicago.

Kuznets, Simon. 1955. Economic Growth and Income Inequality. *American Economic Review* 45: 1–28.

Naghshpour, S. (2012). *Statistics for economics.* New York: Business Experts Press.

Nickerson, D. (1994). Construction of a conservative confidence region from projections of an exact confidence region in multiple linear regression. *The American Statistician 48*(2), 120–124.

U.S. Census Bureau. (2011). *Regions—All races by median and mean income: 1975–2010.* http://www.census.gov/hhes/www/income/data/historical/household/

Zellner, A. (1962). An efficient method of estimating seemingly unrelated regression equations and tests for aggregation bias. *Journal of the American Statistical Association 57*(298), 348–368. http://technet.microsoft.com/en-us/magazine/ff969363(printer).aspx

Index

Announcing the Business Expert Press Digital Library

Concise E-books Business Students Need for Classroom and Research

This book can also be purchased in an e-book collection by your library as

- a one-time purchase,
- that is owned forever,
- allows for simultaneous readers,
- has no restrictions on printing, and
- can be downloaded as PDFs from within the library community.

Our digital library collections are a great solution to beat the rising cost of textbooks. e-books can be loaded into their course management systems or onto student's e-book readers.

The **Business Expert Press** digital libraries are very affordable, with no obligation to buy in future years. For more information, please visit **www.businessexpertpress.com/librarians**. To set up a trial in the United States, please contact **Adam Chesler** at *adam.chesler@businessexpertpress .com* for all other regions, contact **Nicole Lee** at *nicole.lee@igroupnet.com*.

OTHER TITLES IN OUR ECONOMICS AND FINANCE COLLECTION

Collection Editors: **Phil Romero and Jeffrey Edwards**

www.ingramcontent.com/pod-product-compliance
Lightning Source LLC
Chambersburg PA
CBHW071839200326
41519CB00016B/4181